Toxic Work

In the series Labor and Social Change,
edited by Paula Rayman and Carmen Sirianni

TOXIC

Women Workers

 Temple University Press
Philadelphia

WORK

at GTE Lenkurt

Steve Fox

Temple University Press, Philadelphia 19122
Copyright © 1991 by Temple University. All rights reserved
Published 1991
Printed in the United States of America

The paper used in this publication meets the minimum
requirements of American National Standard for Information
Sciences—Permanence of Paper for Printed Library Materials,
ANSI Z39.48-1984 ∞

Library of Congress Cataloging-in-Publication Data

Fox, Steve, 1944–
    Toxic work : women workers at GTE Lenkurt / Steve Fox.
      p.   cm.   —(Labor and social change)
Includes bibliographical references and index.
ISBN 0-87722-816-7 (alk.) :
1. Women electronic industry workers—Diseases—New Mexico
—Case studies.   2. Women electronic industry workers—Health
and hygiene—New Mexico—Case studies.   3. GTE Lenkurt—
Employees—Diseases—New Mexico.   4. GTE Lenkurt—Employees—
Health and hygiene—New Mexico.   I. Title.   II. Series.
HD7269.E372U64   1991
363.17′91—dc20   90-45637

To the workers at GTE Lenkurt,
and to my family,
Niame, Mason, Andir, and Shanti

Contents

Acknowledgments   xi

Introduction   3

1   The First Plaintiff   17

2   GTE's History in Albuquerque   29

3   Health Complaints and the
Deaths of Friends   49

4   Federal Investigations, Riots, and Strikes   65

5   Josephine Rohr Builds a
Case Against GTE   79

6   Medical Experts: Interpreting Facts   99

7   Settlement: Deals for Verdicts   119

8   Suing the Chemical Manufacturers:
Dow, Du Pont, and Shell   137

Conclusions   151

x   Contents

Notes   161

Selected Bibliography   167

Index   173

Acknowledgments

This book would not have been possible without the cooperation of Josephine DeLeon Rohr, who opened her office and her life to me from the start. Her openness carried over to her staff, including Robert Work and Claudia Work, her children. I appreciate their patience with my repeated questions, requests for copies and interpretations, and false starts at understanding the big picture. Becoming a participant-observer in the case required that I be allowed to become a participant-observer in the family as well.

I also owe thanks to Judge Woody Smith for being kind, available, and as open with me as his role allowed him to be. Roger Jatco and Pat Maloney, Sr., were of great help and reassurance at several stages of my research.

I thank my mentors at the University of New Mexico for their critical but flexible advice: Vera Norwood, Susan Tiano, Rob Schwartz, Bill Wiese, and Phil May. Individually and as an interdisciplinary committee, they offered key suggestions that kept me revising and enthusiastic support that kept me going.

I would not have arrived at my final draft without the comments of Karen Reeds, Marc Lappé, Donna Mergler, Rosemarie Bowler, Bob Harrison, Mandy Hawes, and Lenny Siegel. I owe Louise Lamphere thanks for sharing with me her GTE Lenkurt press clippings file.

To all the plaintiffs and other GTE Lenkurt workers and

former workers, I say thank you and congratulations for telling your stories as wonderfully as you did. I especially want to thank Ellen Kayser for opening her heart to me.

I appreciate my family's tolerance over the three years I was obsessed with this case, this book, and the people who speak through its pages. I especially thank my wife, Niame, for her conversation, her instinct for the gut issues, and her steady friendship.

Toxic Work

Introduction

At the Hilton Hotel in Albuquerque in March 1972, city fathers and executives from GTE Lenkurt met for a mutual appreciation dinner. General Telephone and Electric Corporation (GTE) was, and is, one of the world's largest communications empires. Lenkurt, a subsidiary specializing in telephone switching and transmission equipment, originated in Silicon Valley in 1944 and was named for its two founders. After a national search for a good location, the company had relocated many of its major manufacturing operations from Silicon Valley to New Mexico, and the dinner was held to celebrate the completion of a major plant in Albuquerque. (Although I generally use "GTE" to refer to the multinational corporation and "Lenkurt" to refer to the subsidiary, participants in the case and Albuquerque residents often use the names interchangeably.)

Lenkurt president Charlton (Chuck) Hunter, who had made the decision to build a plant in New Mexico, took the microphone. With him on the dais was a select group: a banker, a director of the Industrial Foundation of Albuquerque, a homebuilder serving as chairman of the city commission, and an executive of Sandia National Laboratories, the nuclear and aerospace research and development facility. "The people that attracted us here was . . . you people," Hunter said, gesturing at the city fathers. "The bonus is the people employed at the plant. They are proud, talented and dedicated." [1]

Hunter and his Albuquerque hosts had consummated an important deal for both. For Lenkurt, it was a major investment in manufacturing capability located in an area offering cheap labor and what both sides emphasized was "a good business climate."

Not just a euphemism for antiunionism, the phrase expresses an opposition to welfare, foodstamps, and the minimum wage as well. As Barry Bluestone and Bennett Harrison found in researching their 1982 book, *The Deindustrialization of America: Plant Closings, Community Abandonment, and the Dismantling of Basic Industry*, corporate executives and industrial recruiters, over the last fifteen years, have used "good business climate" to mean a dismantling of the "social wage"—the full range of benefits and services for workers. "The social wage is costly to business," Bluestone and Harrison say, "and increasingly they want out."[2] Bluestone and Harrison found that

> in one recent study by the Conference of State Manufacturers' Associations, the "goodness" of a state's "business climate" was defined in terms of low taxes, low union membership, low workmen's compensation insurance rates, low unemployment benefits per worker, low energy costs, and few days lost because of work stoppages—in that order.[3]

The owners and managers of industry are thus the primary defaulters in the social contract binding corporations, communities, and workers in cooperation and mutual benefit.

For the city's industrial recruiters, the new plant was a trophy in the national race to land high-tech companies. Though the plant was big to Albuquerque, it was only a small part of a global operation. GTE was playing for high stakes in the 1970s. In 1977, for example, GTE chairman and CEO Theodore Brophy predicted at a news conference in London that the number of telephones in the world would double during the coming decade.[4] This growth spurt in telecommunications, he said, was made possible by changeover from mechanical to electronic switching, digital transmission, fiber optics and satellite transmission—innovations that would benefit people the world over, making communication faster, cheaper, more varied, and more reliable. Globally, the telecommunications field then invested about $30 billion annually in new development; Brophy predicted that would double by 1988. Brophy's salary in 1978 was $538,000; GTE's revenues that year were $8 billion, half from communications and half from the manufacture of high-purity chemicals, photographic flashbars, and data transmission equipment.[5]

During the late 1970s and early 1980s, GTE jockeyed for competitive position with AT&T and foreign conglomerates, bidding on giant contracts for overhauling the telephone systems of Egypt, Iran, and Italy. As deregulation of the Bell system caused major shifts in the communications industry in the United States, GTE juggled its subsidiaries and added new talent in management and marketing. *Forbes* magazine reported in 1978 that because of Brophy's efforts to "restructure [the corporation] right from the top to the bottom," GTE was becoming known as a "major corporate bodysnatcher."[6]

To fill the position of corporate president, Brophy hired Tom Vanderslice, who held a Ph.D. in physics and chemistry and had been a Fulbright scholar. In 1981, Vanderslice sounded a call for open communication within GTE:

I can't manage our Salt Lake or Albuquerque operations from here [Stamford, Connecticut], but I can sure get a management structure in place, and staffed with the best kind of people to do the job. . . . Everything works to get you isolated, but you can't let that happen. You have to talk to the drivers, the [executive aircraft] pilots, for example. And, you've got to have people who are not afraid to disagree with you.[7]

This statement, reported by *Industry Week*, sounds like a clarion call for participatory, or at least open, management—themes borrowed from our highly successful Japanese competitors.

However, Brophy's intentions to restructure management "from top to bottom" did not reach the *real* bottom. Far below Brophy and Vanderslice, there were layers of workers vitally necessary to the corporate mission but completely invisible to executives at the top. At this level was the shop floor, where unskilled assemblers, mostly minority women, used solder, epoxy, solvents, acids, plastics, and other toxic chemicals to assemble the solid-state devices inside electronic components. Many workers in this true bottom level took issue with company policies—in fact, they may have been ill and dying from them—but Vanderslice's call for open disagreement apparently did not reach their supervisors or the managers of the Lenkurt division.

In 1986, I discovered some of those voices from the shop floor. Many of them had been disagreeing with GTE Lenkurt managers for nearly fifteen years, and had been able to get the company's attention only by filing claims of occupational disease and disablement caused by overexposure to toxic chemicals. Most of them were Hispanic women, and they were not in the loop of mutual admiration formed by Chuck Hunter and the Albuquerque city fathers, or in the dialog of managerial feedback enjoyed by Ted Brophy and Tom Vanderslice. Those at the top of this far-flung multinational corporation either were not aware of the lowest strata in their own companies, did not think life was different for those workers, or did not care. Perhaps it was a mixture of all three. Furthermore, Vanderslice and Brophy were insulated from the workers' struggle for recognition of their chemical-related disabilities by state laws narrowly defining occupational disease, by limited compensation schemes, by channels within the corporation that assigned responsibility for employee illness and injury to insurance carriers and their risk-management attorneys, and by barriers of class, race, ethnicity, and gender.

The GTE Lenkurt plant in Albuquerque had had a stormy struggle with its production workers throughout the 1970s and early 1980s, beginning soon after the mutual appreciation banquet at the Hilton, as I found in 1986, doing research for my doctoral dissertation in American studies at the University of New Mexico. The strife peaked in widely reported strikes and demonstrations between 1978 and 1981. Then in 1984 the simmering relationship coalesced into a string of occupational disease suits. That year one fired worker, dying of cancer, chanced upon a sympathetic attorney, an Italian-Hispanic woman who had come to the United States as a political refugee from the Dominican Republic some years before. The worker was unemployed and uninsured, and her home was being sought by a hospital to cover her $70,000 medical bill. Helping her led the attorney to many other ill Lenkurt workers. By 1986, lawyer Josephine DeLeon Rohr had filed individual claims of occupational disease and disablement for 85 GTE workers and ex-workers.

In the next two years the plaintiff list would grow past 250, 95 percent female and 70 percent Hispanic. The list included people with cancers not often found in New Mexican Hispanics, frequent miscarriages, excessive menstrual bleeding and hysterectomies, odd neurological problems, and a strange array of other conditions. These workers had the passionate conviction that they and their friends had been well before working at GTE Lenkurt and that something about working there had been dreadfully bad for them. GTE was equally convinced that their occupational histories had nothing to do with their medical problems.

Within fifteen minutes of meeting Josephine Rohr in March 1986 I decided that the whirlwind of issues she was struggling with was a good match for the interdisciplinary American studies methods in which I had been trained. It would prove, in fact, to be a far greater test of a single researcher's point of view than I could have realized.

I found her in an office stacked with boxes of documents supplied by GTE Lenkurt. She had been pestering the corporation for two years for worker employment and health records, chemical documentation, manufacturing processes, and other records necessary to her cases. Working alone or with inexperienced part-time colleagues, and with her college-age son and daughter as staff, Rohr was putting in sixteen-hour days and seven-day weeks. Her demeanor veered from grief at her clients' suffering to rage at bureaucratic malfeasance to black humor at the surreal quality of her situation: battling a multinational alone. Her first client in the case had died four months before my visit, three others had attempted suicide two weeks before, and she had noticed a car that followed her each evening when she left the office at 9:00 P.M.

To understand the case, an observer would have to attempt to coordinate several kinds of information coming from disparate disciplines, many of them technical specialities. In attempting this synthesis, I merely followed Rohr, who, though she was using the specialized procedures of law as the primary means of advancing her clients' interests, was also on a voyage of discovery from week to week.

GTE stonewalled for many months, until the state district court judge began ruling in Rohr's favor and ordering the corporation to supply documents she requested. Understanding what had gone on at the GTE Lenkurt plant demanded information and analysis from engineers, chemists, toxicologists, physicians, electrical production workers and supervisors, medical statisticians, labor organizers, and public health advocates. Most of the key experts for both sides were from Boston, Chicago, Montreal, Silicon Valley, or the San Francisco Bay area. The working-class experience of the plaintiffs had to be translated into the professional-class understanding of the specialists, and barriers of region and discipline were added to those of gender and ethnicity already implicated in the original problems.

Information from one source had to be evaluated through the values of others, and when any technical analysis became too arcane or abstract, the unfolding story of the case continued taking its momentum from the personal struggles and tragedies of the women workers. Rohr was not only a dogged and courageous solo legal practitioner, but also the emotional lightning rod for the domestic and personal dramas of the workers. Like the judge, I became most convinced that the case demanded a full hearing after reading the naive and wrenching depositions of the workers.

By 1987, the list of clients had grown to nearly 150. It was a tough year. Several more clients died. Rohr's office was burglarized and her computer stolen, along with the disks storing the information from 3,000 hours of data entry and analysis (luckily an operator had made backups only the day before).

Rohr moved her office to another location and acquired financial backing and legal help from a prominent San Antonio personal-injury trial attorney. Preparation for what promised to be the biggest occupational disease and disablement trial in New Mexico history was speeding up.

In late May, with trial set for June 15, I was given the opportunity to become a participant-observer. I faced a dilemma: for my dissertation I needed to follow the trial preparations and strategy ever more closely; at the same time, as the stepfather of three teenaged sons in a university town offering little summer work opportunity, I needed to earn some money. Rohr, who had just affiliated with the San Antonio attorney with financial resources, offered me $7 per hour to help round up medical records not yet complete. Neither she, her colleagues, nor my university advisers thought my participant-observation posed any conflicts. All expected that, after nearly three years of pretrial discovery, they would soon be hearing the issues debated in open court. However, the day before the trial was scheduled to start, GTE suddenly delivered a settlement offer with a key contingency: Rohr had to obtain unanimous consent from 115 plaintiffs and their spouses.

The judge agreed to suspend the trial on a day-to-day basis while Rohr sought approval from her clients. This sudden need to assemble more than two hundred people and explain to them, family by family, a settlement offer couched in twelve pages of legalese with strict demands for secrecy and anonymity, threw Rohr's office and staff into a weekend-long marathon of explanation, diplomacy, and family counseling. The plaintiffs were working-class people with little experience with legal proceedings; many had stress disorders that included cognitive and memory limitations suspected to have been chemically caused. Some were unable to easily comprehend the proceedings; others were angry at the dictatorial terms and tendering of the offer. Many had to bring children, and the office complex, while new, was not designed to accommodate dozens of people waiting for hours, let alone on a 98-degree June weekend with the air conditioning on reduced capacity. It was a circus of pressure, with people lining the halls and stairwells, sitting on the floor. It was my first chance to meet many of the plaintiffs in person and see their understanding and reactions to the case. For that and many other reasons, I felt compelled to assist in the effort to explain the settlement offer to the beleaguered plaintiffs.

The settlement offer included a clause barring Rohr or any of her employees from discussing the terms of the settlement in public. Although I had worked for Rohr and her colleagues for only four weeks, I fell under the gag order. Luckily, the only information of importance I had gained while an actual employee of Rohr's was the exact dollar figure of the settlement and the details of each worker's share. Having earlier studied the preceding settlement proposals by both sides, I knew the approximate figures and terms anyway. In May GTE's defense attorney, Carlos Martinez, had given me clearance to publish anything I had learned before I became Rohr's employee. That included 99 percent of what I knew about this massive controversy.

High-tech Workers, Health, and the Law

The basic assembly work of electronics is done around the globe by women working for pennies an hour. The "clean industry" may not have smokestacks or visible piles of scrap, but it is an industry based on chemicals with lethal effects. The spotless, solid-state, steel and ceramic innards of our electronic tools and toys have at every step in their assembly been dipped into grease-dissolving solvents; etched with acids or caustics; plated with salts of heavy metals; coated with plastic films, inks, or emulsions; and glued, stabilized, or sealed with epoxies. Plant design often interferes with removal of the fumes generated by these processes.

Scores of the chemicals used in electronics are known to cause, when inhaled or absorbed to excess, drunkenness, nerve damage, hormonal disruption, heart rhythm problems, kidney, lung, or thyroid damage, cancers, mental dysfunction, and emotional disturbance. Tested quickly by their manufacturers on a few healthy male volunteers, they are sold at great profit and put into production processes around the world before anyone knows what their effects will be on workers of various sizes and health conditions, on women, or on workers under the stresses of speedups and overtime. The dirtier plants, like GTE Lenkurt, can expose workers to synergistic mixtures of so many different chemical fumes that pinpointing causes of symptoms is impossible. Effects on workers may be acute and sudden, or cumulative and latent.

The GTE Lenkurt case is the largest example of these problems yet reported in the U.S. electronics industry and the largest workers' compensation case in New Mexico history. As this story reveals, workers suffering such conditions face not only the ignorance and skepticism of a medical community untrained in toxicology, but a system of disability laws and compensation schemes designed to award a fraction of a living wage to a fraction of those injured. Only the cases lucky enough to encounter a Josephine Rohr even get so far as being filed properly, and those that threaten to get to trial are usually settled out of court—as was this one, under terms forbidding the workers from telling their stories to the public or divulging the amount they were paid.

Out-of-court settlements usually also require destruction of all documents generated by the plaintiffs, and forbid expert witnesses from using in scholarly publications the data gathered in studying the plaintiffs. In other words, a settlement completely privatizes the relationship between worker and employer, no matter how legally or medically significant for others, and no matter how serious the impact may be on the life of the community. Such settlements are a form of white-collar hush money.

This book is the sole public record of six years of legal and medical investigation into the lives of some 250 high-tech workers. (Several studies have been published by occupational health researchers who volunteered their time to study the plaintiffs during 1988 and 1989; their articles are cited here.) The book includes the crucial testimony of expert medical witnesses for both sides, revealing the role of personality and ideology on professional testimony, and the distortions that professional testimony can make of plaintiffs' stories. The book is based on 100,000 pages of pretrial discovery documents and hundreds of hours of interviews. GTE corporate executives and attorneys decided to prohibit all nonplaintiff employees from giving interviews except under deposition.

Although it incorporates a history of the plant and some analysis of work pace and process, this book is largely the story of the way the women workers and their attorney had to improvise in their struggle to gain information and justice. It is a story of chance encounters. It is a story of the power of personality and position in corporate life and in local, state, and national public health and labor agencies. It is thus a study not only of one controversial company, but also of the archaic, random, narrowly focused, and penurious set of institutions—not a real system—that many injured workers must rely on for support. It is also a window on some of the consequences of the atmosphere surrounding high-tech development in the last twenty years: the uncritical enthusiasm, the unconditional belief in high tech as a panacea, and the futuristic optimism shared by political and industrial power brokers as they look toward the prophesied postindustrial information age. The story of the GTE Lenkurt workers supports the argument made in recent books critical of unbounded faith in microelectronics manufacturing.

Lenny Siegel and John Markoff's *The High Cost of High Tech: The Dark Side of the Chip* and Dennis Hayes's *Behind the Silicon Curtain: The Seductions of Work in a Lonely Era* portray electronics and microelectronics manufacturing as industries with technical and social problems clearly affecting workers adversely at many levels of the industry.[8] Clean buildings may house minority women doing most of the basic assembly steps, trapped in a second society of meaningless, low-paying work with significant hazards from a long list of toxic chemicals. Filtration systems may remove particles but not fumes. The furious pace of innovation, while a source of profits and a sense of progress, may erode the personal and family strength it is purported to support, by elevating work to a controlling force in people's lives. Members of the Macintosh development team at Apple in the early 1980s, for example, wore T-shirts proclaiming, "Working 90 hours a week and loving every minute of it!"[9]

Robert Howard, author of another book critical of recent corporate life, *Brave New Workplace*, urges us not to confuse the freedom and convenience of consumer products with the identity of the industry that produces them. IBM's advertising in recent years has used the images of Charlie Chaplin's Little Tramp and the familiar actors who starred in MASH. Using the beloved figure who got caught in the gears of industry in *Modern Times* and the caring members of the quirky MASH family is an obvious appeal for us to set aside our fear of machines and their power over our lives. The truth behind these image manipulations, Howard reminds us,

> is that work in America is essentially a relationship of unequal power, that conflicts of interest are endemic in working life, and that this new model of the corporation, like the old, is founded on systematic denial of influence and control to the large majority of working Americans. . . . Thus, when technology is linked to the imperatives of corporate control, work often becomes the antithesis to the realm of freedom that the image of Charlie Chaplin before the personal computer suggests.[10]

The arguments between the boosters and the critics of the high-tech corporate world are really the latest expressions of two conflicting conceptions about the place of work in society. Capitalism has always seen work as a private contract between individual entities, with regulation of work best left to private corporations. Workers, protesters, and reformers have often held an opposing conception of work as a public activity far too important to our collective life to be left to private business interests. Howard points out that if we make the corporation the dominant institution in our lives, we will do so only at the price of weakening the other institutions that have historically mitigated the worst effects of the private business control of workers: the industrial union movements of past decades, public health programs, and agencies like the Occupational Safety and Health Administration (OSHA), the National Institutes of Occupational Safety and Health (NIOSH), and the Environmental Protection Agency (EPA). If we continue to believe the simplistic equation that electronics manufacturing—or any single form of corporate activity— means an end to exploitation, pollution, and meaninglessness, we will impoverish our culture by weakening the diversity of political interests that is our sole guarantee of pluralistic and democratic life.

This story of a stubborn and empathetic woman attorney and more than two hundred workers in a plant that was the pride of a city shows how an ill wind can blow through the clean corridors of the high-tech workplace, changing the workplace, the lives of the workers, and the very meaning of work.

The First Plaintiff

In August 1984, a young woman walked into the Albuquerque law office of Josephine DeLeon Rohr. Rohr, though only two years out of law school, did not need walk-in clients. Contracted to the local office of a national firm, she was making very good money for the first time in her life. She specialized in collections, personal injury, and workers' compensation, and her income of some $8,000 to $10,000 a month represented great stability and insulation for her.

In 1955, at the age of sixteen, Rohr had fled her native Dominican Republic after her two best friends had been murdered by dictator Rafael Trujillo for their participation in a student antidictatorship group. She moved to California, had three children before she was twenty-one, then was divorced from their father and supported her family as a bilingual paralegal in a rural legal assistance program. Her second husband died in Vietnam, and her oldest son died suddenly of leukemia at eighteen as she was beginning law school in 1978. By 1984, after what seemed a lifetime of scrambling and loss, Josephine Rohr was looking forward to some peace and stability.

The woman who walked in her office that day in August—Amy Cordova Romero—was in desperate crisis. Rohr at first took her for a teenager, but she was actually thirty-seven and struggling through chemotherapy and complications associated with metasticized ovarian cancer. Divorced, jobless, and with no health insurance, she owed St. Joseph Hospital about $70,000, and the hospital was suing her for title to the fourplex in which she and her teenaged son lived. Deeded to her in a divorce settlement, it was her only property and source of income. Amy's life was dwindling on all fronts.[1] Her parents were in their eighties and near death, and her ex-husband paid child support only erratically. It seemed a pathetic situation, and clearly there was no possibility of paying for legal counsel. Looking across at Amy, Josephine was struck by her pretty features and unlined skin. "You know," she said, "you sure don't look like the stereotype we have of the terminal cancer patient. You seem so young."

"I know. But plenty of the girls I worked with have been sick and lost their uteruses and stuff, and they were as young as me. Some were younger."

"Plenty of them?" asked Rohr. "How many did you work with?"

"About twenty-five or thirty," Amy answered. "All of them had hysterectomies."

"All of them? That's impossible!" said Rohr. "Are you serious? Who are they?"

Amy began naming names, and Rohr scribbled them down. When she had named about ten women, Amy said,

> There are more. I can't think of them all right now. We worked at GTE Lenkurt making electronics things. I worked there for most of the 1970s. There were all these chemicals we were always smelling or getting on ourselves. We used to get sick a lot and have real heavy periods, but we thought it was just us. But since I've been in the hospital, I've heard that nearly everybody we worked with has had operations, mostly hysterectomies, and lots have had cancer.[2]

Rohr assured Amy she would help her delay the hospital's seizure of her house and told her not to worry about paying her. After Amy left, Rohr stared at the list of names and wondered if it could possibly be true. She called some of the phone numbers, and yes, the women had had hysterectomies while working with Amy. They had had other conditions, too, and they gave her more telephone numbers, and yes, those other women had lost uteruses, or thyroids, or the ability to think clearly or breathe easily, or some other key element of their formerly healthy bodies. And there were a number of reported cancers. All while working at GTE Lenkurt, making electronics components and "things."

Word got around, and women started calling Rohr's office. Before she had even written the hospital's attorney on Amy's behalf, she had notepads filled with lists—human organs, ill women with Hispanic names, and chemicals that sounded oddly familiar, even feminine: ethylene . . . urethane . . . methyl something . . . "trichlor," or something that sounded like bleach. She forced herself to focus on Amy's immediate crisis with the hospital, but curiosity about electronics work crowded in. She tried to imagine the factory production line: where would Amy and her friends be, what would they be doing, what would the chemicals smell like? She could not stop herself from calling more names, and each new name added to her amazement.

Thus began Josephine Rohr's chance investigation into the life and health of Amy Cordova Romero and 250 other former GTE workers, nearly all women and mostly Hispanic. A novice who had never tried a case, Josephine Rohr was taking on one of the top twenty multinational corporations in the United States with defiance, naiveté, and moral outrage. Mortgaging her house, borrowing money, improvising "like a bull in a china shop," as she described herself, clinging to faith in the system, and finding activist allies when she needed them most, Rohr battled GTE for nearly three years. Her campaign uncovered ruthless management, a toxic arsenal of electronics production chemicals, and a stratified work force where minority women did meaningless, repetitive work under dangerously contaminated conditions.

Amy Cordova Romero had asked three lawyers to help her delay the hospital's seizure of her house, but all had refused. Desperate, she asked her pastor if he could recommend someone. He had heard that Rohr was active in evangelical Christian work; she also did a legal information show on a local Spanish-language radio station. The day after Rohr accepted Amy's case, a fellow lawyer asked her, "Why would you waste time on a lost cause like that?" Rohr replied, "It's called compassion."

Rohr's Caribbean Teenage Political Education

Josephine DeLeon Rohr was born in the Dominican Republic, one of two daughters in a wealthy and powerful Italian-Dominican family. Her mother, Teolinda Bencosme, was the eldest daughter of Cipriano Bencosme, who was elected president of the country on November 25, 1930, and assassinated five days later by the dictator Rafael Trujillo, who ruled the country for the next thirty-one years with U.S. backing.[3] U.S. Marines had occupied the Dominican Republic from 1924 to 1930, and Cipriano Bencosme's campaign emphasized getting the Marines out. Several Bencosme men subsequently served as senators and provincial governors as the family went in and out of favor at Trujillo's whim. Sergio Bencosme, Rohr's granduncle, who served as minister of foreign relations in the administration preceding Cipriano Bencosme's election, was assassinated in New York City in 1935, possibly by gunmen mistaking him for another figure. Trujillo himself was killed by two of Josephine's distant cousins in 1961.

Josephine's parents were wealthy enough for her father to lose $100,000 in weekend poker games with his wife's brothers. The Bencosme wealth came from coffee plantations and other landholdings. They owned entire villages of tenant workers. Rohr says her sharpest memories include the empathy she felt watching workers queue up for free commodities her mother handed out weekly from a table in front of the gate to their house. "It made me feel horrible for them. Why were things set up so they could not make a dignified living for themselves?"

Josephine's mother had strong ideas about the value and inaccessibility of education. She looked for talented children in servant and peasant families, paying for their college and paying the parents the lost value of the children's labor. Josephine claims that this patronage resulted in twenty to thirty physicians and lawyers now practicing on the island. Josephine herself skipped from the third to the seventh grade, finished high school at twelve, and went straight into law school. She was a striking, precocious law student of fifteen when she became tangled in Trujillo's intrigue.

Chaperoned by her godparent Francis Townsend, an attaché at the American Embassy, Josephine attended a party at the embassy in 1955 for Porfirio Rubirosa, whose companion was Zsa-Zsa Gabor. Townsend introduced her to Rubirosa, who was a close friend of Rafael Trujillo, Jr., the dictator's son. Within a few days, the elder Trujillo, then sixty-three, began sending cars for Josephine and directed one of her uncles to install her in one of Trujillo's houses. The uncle complied, but Josephine's mother heard of the arrangement and told Trujillo, "Go to hell." Furious, Trujillo investigated Josephine and her friends and discovered that she and two high school classmates had become involved with a revolutionary anti-dictatorship group. Trujillo had the two friends, Minerva and Theresa Mirabal, delivered to jail, where they were raped by the male prisoners and killed in front of their boyfriend and husband.[4] The Bencosmes had by this time escaped the country. They sent Josephine to Central America with an Italian-Argentinian couple, giving them $20,000 in silver and instructions to see that she joined another godparent who was a professor of medicine at UCLA.

Refused sanctuary with the UCLA relative, Josephine made her way to San Francisco, where she used her affidavit of support from the archbishop of Honduras to obtain a work contract with a private school. She began learning English and resumed her studies. She stayed with relatives of the American diplomat Averill Harriman, then moved to a boarding house where she met and married a law graduate from Ohio State University. He was thirty-two, she seventeen. Four years later, after having two boys and a girl, she divorced him. Twenty-one, with three toddlers and poor command of English, Josephine parlayed her Dominican Republic law school experience and fluency in Spanish into a job as legal assistant to Cruz Reynoso, an attorney who later became a justice of the California supreme court.

Within a year, she became one of the first paralegals hired by the new California Rural Legal Assistance (CRLA) program. Given her choice of locations, she chose to run the CRLA office in Salinas, where, with Chicano farmworkers as her clients, she was allowed to "practice" law without a degree before the workers' compensation board, insurance board, and immigration and Social Security offices. During the time Josephine ran the Salinas staff, the federal War on Poverty program named the CLRA program the best legal assistance service in the United States for four years running.

Rohr's second marriage, to army lieutenant John Rohr, ended after three years with his death in Vietnam. After two years in New York with her mother and sister, she decided to relocate to a warm climate with Hispanic population, and chose Albuquerque because a residential community that advertised heavily in New York media, Rio Rancho, offered her a week's stay and tour of the area for $350. In 1974, at thirty-four, she enrolled in undergraduate classes at the University of New Mexico. In 1978 she was accepted by the University of New Mexico Law School. Ten days before she was to begin classes, her oldest child, Alan, died suddenly of leukemia at eighteen. Rohr went ahead with law school but suffered a delayed breakdown, taking a year and a half off before her final year to seek counseling for her grief and the grief and confusion of her middle child, Rob, then sixteen.

And so, when Amy Cordova Romero called at her office in 1984, Josephine Rohr was not the typical recent law graduate. Raised in a violent and misogynist Caribbean society, she had a well-developed hatred for abuse of power and oppression of women. She was a feminist and populist without any particular ideology beyond that embodied in a hand-lettered desk card quoting Abraham Lincoln: "Have faith that right makes might." She had been a victim and a survivor, an insider and an outsider, a patrician and a refugee. For an unemployed Hispanic woman in health crisis and financial trouble, like Amy Cordova Romero, Rohr had more empathy to offer than most lawyers.

Amy Cordova Romero's History

At the time of her initial interview with Rohr, Cordova Romero had been off the GTE payroll for five years. According to her deposition and GTE personnel file, she started work at the plant just after it opened in 1971, at the age of twenty-four. She was married, with a three-year-old son. She had previously worked as a school aide in Santa Rosa, New Mexico, her birthplace. Amy made $1.61 per hour when she was hired in 1971, and $3.90 per hour when she was fired eight years and nine months later. Struggling through a divorce and a severe case of pneumonia, she had asked GTE for a leave of absence. Two months after she returned, her supervisor discovered that Amy had taken credit for some incentive piecework she had not actually finished. She was fired by telegram from Lenkurt offices in San Carlos, California, that same day. With skills not transferable to any other business in the state and a son ten years old, her standard of living deteriorated over the next four years.

In November 1983, Amy got what she thought was a cold, but it did not go away. Her face, arms, and legs became bloated. She had no personal physician, no health insurance, no funds. After a week of having difficulty with breathing, she had her boyfriend take her to the emergency room of the county hospital, operated by the University of New Mexico. She explained her breathlessness, her bloating and weight gain, a "jumping" pain in her chest, chills, and a floating sensation. The emergency room physician on duty diagnosed her primary condition as constipation. When she returned for a followup one week later, the physician told her he had given her the wrong pills for constipation the week before, and this time prescribed laxatives. That Christmas she was sluggish and barely able to observe the holidays. Her condition waxed and waned over the next three months. In March, it became quite grave:

> It got to the point, finally, that I was like choking. I couldn't take it any more, so I finally told my son and my boyfriend that . . . I was dying. I couldn't eat nothing for weeks, and I choked if I laid flat. I was completely out of breath. That's when I told them, "Do whatever you want." So they called my sister-in-law and she took me in to the emergency room at St. Joseph's. I couldn't lie flat for the X-rays, so they brought some kind of portable scanner thing into the intensive care room where I was. None of the doctors could figure out what was wrong with me. This Orien-tal doctor got the portable scanner going and suddenly he gasped and ran to the phone. He called the other doctors and said, "I found what's wrong! Or part of what's wrong!"

The scanner had located the source of the pressure on Amy's breathing: malignant pericardium, a heart sac attacked by cancer cells and filled with blood. In swift succession they found cancer in the left lung and abdomen, spread from its source in the ovaries. The doctors found her pelvic area to be "one rock-hard mass, sidewall to sidewall." On top of that, she had severe pneumonia.

After treating her hemopericardium and pneumonia, doc-tors diagnosed chronic, restrictive pulmonary disease. She had never smoked and had no family history of lung problems. She underwent a hysterectomy with removal of her ovaries and the omentum, a portion of the abdominal cavity.

Chemotherapy began; nurse notes in her medical record report visits to Amy's room by her boyfriend, with alcohol on his breath, who urged her to refuse chemotherapy and "go for the gusto." At home, he fought with her and with her teenaged son. Amy's bills mounted. She was admitted to the hospital on twenty-one occasions between March 19, 1984, and January 29, 1985, including a thirty-one day stay in the intensive care unit. She had been a late-in-life baby, and her parents were in need of care themselves. Within the year they died in Santa Rosa at eighty-three and eighty-four.

While Amy was in the hospital, Patsy Aragon, one of her former co-workers and best friends from GTE, came to visit and told her that many other women they worked with had been having problems; many had had operations, mostly hysterectomies. In fact, Patsy said, she herself and one other girl seemed to be the only ones from their department at GTE that *hadn't* had operations.

Released from the hospital after the major surgery, Amy resumed trying to manage her fourplex. "Is that how you supported yourself?" asked GTE's attorney when taking her deposition a year later. "Well, I sort of supported myself," she replied. The hospital billing department urged her to go to the American Cancer Society to ask for help with her bills, but they were of no help. When the papers came announcing St. Joseph's intentions to ask the courts to seize her fourplex, Amy realized she had no chance without a lawyer. In the initial interview with Amy in 1984, Josephine Rohr did some calculations: $600 total rental income from three units, plus $200 irregular child support, minus the $523 mortgage payment and minimum housing maintenance, utilities, gas, and oil for her 1980 Honda, prescription drugs, $20 for her son's monthly allowance, and car insurance, left $93 per month to provide food, clothes, entertainment, and emergency funds for two people. Amy had accumulated fifty-seven shares of

GTE stock, which paid her a total dividend in 1984 of $47.65. Her boyfriend occasionally contributed toward necessities.

Josephine Rohr responded viscerally to Amy Cordova Romero's stoicism. An intricate pattern of coincidence contributed to their rapport: shared ethnicity, gender, divorce, parenthood, cancer, and family loss. Had Amy never found Rohr, perhaps none of the other plaintiffs would have found help either. Had Rohr not given Amy the opportunity to mention the others, she might never have heard of them. Had she not called them and confirmed their illnesses, she might never have believed Amy's condition to be work related. Without the work relatedness, the only question before Rohr would have been what Amy owed the hospital, instead of what GTE owed many workers.

GTE's History in Albuquerque

2 GTE Lenkurt came to Albuquerque in 1969 from San Carlos, California, in the northern reaches of the high-tech region known as Silicon Valley. It was the second of three major electronics manufacturers to locate in the city in response to an intensive recruiting campaign, and leaders in business and local government hailed the firm's arrival with euphoria. Lenkurt executives promised to hire hundreds of New Mexican workers, and predicted rapid growth of the work force and plant budget. True to these predictions, during the 1970s GTE Lenkurt became the largest manufacturing employer in the state, at its height providing up to two thousand workers, mostly minority women, with good paychecks from abundant overtime doing relatively simple jobs.

However, the plant's working environment, according to many employees, was stressful and toxic. The plant looked modern and, with its absence of smokestacks and exterior noise, appeared to exemplify the reputation of electronics as a clean industry. However, worker health records, from company and private sources, reveal puzzling levels of symptoms and illness. The workers' accounts of their experiences at GTE Lenkurt reveal that the plant's modern exterior concealed oppressive management and chemical overexposures.

Albuquerque and GTE Lenkurt Recruit Each Other

In the summer of 1969, the president of GTE Lenkurt, Charlton Hunter, using the alias "Chuck Green" to keep his company's identity secret and avoid premature press speculation, met with Albuquerque leaders to discuss building a major facility in the city.[1] At that same time, city leaders were also wooing Singer-Friden, makers of calculators, cash registers, and adding machines, and the city's negotiators tried to keep GTE and Singer ignorant of each other's presence in town. They wanted each potential recruit to feel like the big fish in the new pond.

Due largely to defense spending by the federal government, Albuquerque's population, only 35,000 when World War II broke out, had nearly tripled during the 1940s, more than doubled again during the 1950s, and continued to grow rapidly in the 1960s. In December 1966, the Atomic Energy Commission phased out a prime contractor operation, American Car and Foundry Industries (ACF). This loss of hundreds of jobs and millions in budget spurred business leaders to create Albuquerque Industrial Development Service (AIDS) and the Industrial Foundation of Albuquerque (IFA) to attract new firms. Within three years, three of the top fifty corporations on the *Fortune* 500 list announced plans to locate facilities in the city: General Electric would build jet engines in the old ACF building; Singer-Friden would build "smart" cash registers and adding machines; and GTE Lenkurt would build telephone switching equipment and other communications components.

One of the primary reasons these firms were attracted to Albuquerque was the city's weak tradition of organized labor.

Singer-Friden and GTE became interested in Albuquerque through Fantus of Chicago, the largest factory relocation service in the world. James Garvin, active in New Mexico industrial development from the late 1960s to the 1980s, claims that Fantus has always had an interest in helping corporations escape the influence of unions. In the late 1970s Garvin headed the state industrial development commission and recommended that the state hire Fantus. Garvin told me in an interview that at that time Fantus revealed that 50 percent of its clients refused to even consider a state that did not have a right-to-work law outlawing compulsory union membership at any plant. New Mexico has never had such a law, but it has had little union activity. Escaping union interference was certainly an issue for GTE when they used Fantus in 1969. "It wasn't just money [higher union wages]," Garvin told me, "it was [union] interference in management prerogatives" that GTE wished to avoid by coming to New Mexico.

In concurring with that point, chief Albuquerque negotiator Ray Powell adds that Lenkurt CEO Chuck Hunter's personality was on a collision course with organized labor. In what the other participants call the key negotiating session, Hunter and Powell met alone. Hunter wanted to be reassured that GTE's competitor, the Bell system, represented in town by Mountain Bell and Sandia Labs, had not "taken over the town." Hunter also wanted to know why, if Albuquerque was such a good place, more major companies were not already here. Powell replied that Sandia and Mountain Bell had benevolent feelings of responsibility toward the community, not proprietary interests, and that GTE should come in and share that responsibility. Having convinced Hunter to build a plant here, Powell nevertheless came away from the head-to-head session considering Hunter "a blunt, tough, bull-of-the-woods industrialist." Says Powell, "I think they were having labor problems over in San Carlos, and I think he was such a bull-of-the-woods that his management practices practically assured [his plant's] being organized here." Garvin had a heart attack during the talks and wound up at the same hospital as two GTE men who suffered gastric attacks. All three were treated by Garvin's physician, who, against his own better judgment, relayed to Garvin GTE's opinions of the negotiations.

The city offered GTE, as it offered Singer-Friden and later Signetics, Digital, and Motorola, interest-free and real estate tax–exempt industrial revenue bonds. GTE put up $2 million and the city put up $4 million to build the 200,000-square-foot plant, and the IFA spent $200,000 preparing the site and bringing in utilities and access roads. As part of the agreement, local press stories reported, GTE pledged to construct a building whose appearance would "conform to the city's architectural traditions . . . for both the present and the future." In an area dominated by the Spanish-Pueblo style, "architectural traditions" was understood to mean that exhaust vents should not extend above the parapet wall surrounding the roof, so as to make the plant look as little like a factory as possible.

The plant opened in 1971. On a pre-opening press tour, Lenkurt president Chuck Hunter and vice president and general manager Thomas Wortman told reporters that the plant was designed and intended to be a complete manufacturing operation which will "begin with raw materials and progress through various stages . . . manufacturing such components as crystal filters, inductors, transformers, thick film microwave circuits, single and double-sided printed circuit boards, and plastic and sheet metal parts." The material specifications, work methods, safety precautions, and precisely the same process specifications were transferred intact from San Carlos to Albuquerque.

GTE had agreed to hire the unemployed and underemployed, training up to 900 at the Albuquerque Technical-Vocational Institute in basic electronics and "Lenkurt's way of doing things." With a few weeks' practice in soldering and other simple assembly techniques, the new employees eagerly began work. There was excitement in the air for these new hires; the press and city fathers were hailing Lenkurt as the flagship of a new armada of high-tech employers. Few people knew what "high-tech" meant, exactly, but it was clearly something modern and good. For beginning production workers, pay was low, even by Albuquerque standards. GTE personnel officials promised there would be rapid advancement for those who learned well, plenty of overtime when production began to hit its stride, and perhaps incentive systems to reward the most productive workers.

Many workers came to Lenkurt from menial jobs. Mary Sena was on welfare after a divorce; her only job experience had been at a local fast-food restaurant and as a maid at a hospital. After a caseworker referred her to the Skills Center at the Technical-Vocational Institute, she secured jobs at an Ampex plant and then Lenkurt. Ellen Kayser once toured the West Coast as a singer with a country and western band, but at the age of forty-seven, after a divorce, she decided to take the Skills Center training. She excelled in Lenkurt's mandatory introductory tools-handling course and "became a workaholic. I did enjoy working there." Yolanda Lozano, who years later would become a spokesperson for disabled Lenkurt workers, recalled that she and her friends loved their early years at Lenkurt because "it paid good money, was easy to learn, and we liked the place except for the smells."

Union organizing began at Lenkurt within the plant's first

year. An organizer for the International Brotherhood of Electrical Workers (IBEW), the union that represented Lenkurt employees in San Carlos, came to Albuquerque and began talking to workers during 1972. Lenkurt management responded by denying workers the opportunity to discuss the union. Their techniques of intimidation included blacklists of union sympathizers, rules against discussing the union on GTE property, written reprimands and warnings, and punitive firings and transfers. The employees and the IBEW organizer appealed to the National Labor Relations Board (NLRB), which held hearings in Albuquerque during July and August. Union supporters charged that by November 1972, fifteen workers, ten of them Hispanic, had been fired for no discernible cause except advocating unionization. The NLRB found GTE in violation of collective bargaining law for the fifteen firings and associated coercive tactics.[2]

By June 1973, a majority of workers had signed cards favoring a union chapter. Lenkurt management appealed to the NLRB to invalidate the worker votes, but the NLRB certified the election and denied Lenkurt's objections, citing that Lenkurt had repeatedly refused to bargain. In July 1975, the NLRB again had to order GTE to recognize the union. Lenkurt management finally signed a three-year contract with the IBEW in 1975. With this capitulation, the GTE Lenkurt plant joined the tiny minority of U.S. electronics businesses that are unionized. As of 1986, only 90 of the 1,900 firms belonging to the American Electronics Association had unions representing their production workers; only 5 to 8 percent of all U.S. high-tech industry workers are members of unions.[3]

Division of Labor and Organization of Work in Electronics

Electronics manufacturing is highly stratified; because of the furious pace of innovation endemic to all branches of electronics, there tends to be a higher percentage of upper-level technicians, engineers, and programmers than in older industries, sometimes reaching 25 percent of the work force. There is a gulf between these technical design people and the production workers. In the majority of Silicon Valley firms, 70 to 90 percent of the production workers are female, about 50 percent of them nonwhite. The percentages were even higher at the Lenkurt plant in Albuquerque— 95 percent of those in the assembly operations were women, and the production work force, males included, was at least three-quarters minority, mostly Hispanic but including Native Americans, Asians, and blacks.

Because of the rapid changes in designs as products were modified and refined, the Lenkurt production operations favored manual assembly over automation—people can be retrained more quickly and cheaply than automatic-assembly machines can be redesigned and rebuilt. Together, the forces of rapid innovation and manual assembly deskill the work, segmenting it into simple steps demanding no comprehension of the overall purpose of the component or understanding of electronics in general. The tasks can be quickly resequenced, eliminated, or modified by engineer-managers as products are redesigned.

Dan Winkless, who worked at the plant as a temporary computer programmer for several months in the mid-1970s, had unusual access to the stockroom and the two production lines: he was hired to write programs specifying the most efficient sequence for feeding parts into machines that manufactured circuit boards.

> It was a very tightly controlled building, like a military base. I wore a tie and had a cubicle in the air-conditioned office area, and even there, there was bad morale and people watched the clock, taking breaks on rigid schedules. There was a time-and-motion consultant harassing the women on the line about the best sequence of reaching, soldering, and so on. I remember one supervisor insisting a woman worker learn to use her left hand to do something, and driving the woman to tears. I had a moral dilemma about the mechanization of people. I could not imagine their being bright people, inside—stopping when the bell rang, having a smoke, going to the powder room, starting up again. It seemed awfully demeaning.

The production work force at Lenkurt's new facility was divided into departments, each under a foreman, with supervisors in charge of smaller divisions. The plant ran two shifts at first, but added a third in the later 1970s. Except for three enclosed shops for the printed circuit lab—where extensive metal plating and acid etching were done—and for the crystal filter and chip capacitor operations, all other departments were housed in one huge area, sometimes divided by partitions five or six feet high. The ceiling was fourteen feet high. It was one of the last "open plants" built in the United States for such electronics work.

While workers were always hired into a specific department, they were often borrowed by other departments as production contracts demanded increased output in some areas and cutbacks in others. The workers' employment records, as reported in their depositions, reveal that each plaintiff was officially assigned over her career at the plant to an average of three departments. Within each department there could be a wide variety of processes in use simultaneously, and processes were changed at a given work station often, according to the deposition of Jack Lacy, a former OSHA inspector who served as safety and loss control officer at the plant from 1979 to 1984.

Department 320—Focus of Work Improvisation

Many of the workers who later became plaintiffs worked in Department 320, Component Assembly, which grew from a handful to over one hundred workers. Most workers, beginning in the early 1970s, spent their time doing repetitive tasks in the making of transformers: one operator wrapped wire around a core "bobbin" in a machine; another stacked E-shaped metal layers together and inserted their middle protrusions into the wire-wrapped core; another dipped the leads of these transformers into pots of boiling flux and solder; another filled the spaces in the core with "potting" epoxy to keep core and laminations stabilized and sealed; another dipped the finished component in varnish or melted plastic. The women in 320 spent most of their time "winding," "stacking," "potting," "snagging," "clamping," "waxing," or "winding daisies," making loops of wire that looked like flowers when spread open to dip in liquid plastic sealant. The tasks were simple enough that most workers did many jobs in each department.

The women in Department 320 used some familiar implements when handling chemicals: Milk cartons, spatulas,

Q-tips, baby food jars, loaf pans, clothespins, and other common household tools were used to hold or apply acids, epoxies, inks, solvents, plastics, varnishes, and so on. Some of these uses were original with the processes brought from San Carlos and some were improvised by workers when Lenkurt began modifying the original processes and designing new ones.

In the same general area of the component assembly department, other workers mixed epoxy cement in cone-shaped paper cups or handmade cones of foil, stirring the mixture with pencils or toothpicks. Lenkurt had altered the formula of an epichlorohydrin-based epoxy to speed its setting time from hours to minutes. GTE engineers in San Carlos decided to mix a resin from Shell with a hardener from the Hysol Corporation, or vice-versa, to eliminate bubbles and speed the setting, against the written warnings of both manufacturers. They called the new quick-setting glue "Lencast." The combination of epichlorohydrin, aliphatic amines, and diglycidyl ethers causes asthmalike breathing problems, eye damage, skin sensitization; also, these chemicals are suspected carcinogens.[4] Fresh batches had to be mixed every ten to fifteen minutes because the accelerated chemical reaction caused it to burn up; smoke rose from paper cups discarded in trash cans beside the workers. There was no special exhaust for these areas; certain work tables had exhaust hoods, but workers report they were always clogged with sticky residue, hair, and dust, so the fumes were free to mix over the entire area.

John Randall, brother of the writer Margaret Randall, went to Lenkurt in 1975 with the covert purpose of helping with union organizing. Randall was never involved with Josephine Rohr's cases, nor did he experience illness or injury on the job. Concealing his college education, Randall got a job in the printed circuit (PC) lab "wet area," where circuits were etched

on coated fiberglass circuit boards in a series of immersions
in rows of large tanks filled with acids, caustics, solvents, and
salts of gold, silver, tin, copper, stannous fluoride, and other
metals. GTE considered this area too dangerous for women
because of the lifting involved, so it was one of the few production areas dominated by men. Randall quickly was promoted
to materials handler (a person who delivers tools, supplies, and
chemicals to other workers) and became the union steward in
the PC lab.

> I delivered all sorts of chemicals to the PC lab, and to other
> areas when they needed me. My supervisors were oblivi-
> ous to any special safety concerns with chemicals. Lots
> of times I saw drums and containers of chemicals marked
> "Warning: Contents Not Tested for Effects on Humans."
> Once I was sent to the stock area to get some stuff, and it
> turned out to be an open barrel of a white, crystalline sub-
> stance. My supervisor didn't tell me what it was. When I
> stood over the drum to get my hand truck under it, I felt
> my whole body just heave and spasm for air. It was invol-
> untary—I couldn't control it. I got really woozy and pretty
> scared that nobody had told me anything about this stuff.
> [*Did you report it?*] Yeah, as the shop steward, I reported
> anything like that to my supervisors. My foreman would
> really get angry and weird when anyone complained. I
> told him about nearly passing out over the barrel, and he
> said, "So?"

According to Randall and several others, the PC lab workers
fought for clothing allowances, since the hydrochloric, sulfu-
ric, and nitric acids they used made holes in their clothes and
shoes. Management eventually added the price of a pair of

blue jeans per week and a pair of shoes a month to their wages while insisting that "nothing here can hurt you."

Janet Caudill worked in Department 345, where crystal filters were made. One day a several-gallon jar of acid broke in a cupboard, sending a yellow cloud into the area. A supervisor yelled that the workers were not to leave the area or they would be "written up," given file memos for disobeying orders. Grace Wessel, also in the department, etched crystals in Corning Ware baking pans full of a yellow acid. It was heated when used, and precipitated into flakes when unheated. Wessel was never told what the acid was. It ate through the Corning Ware periodically, and Wessel says supervisors showed irritation when asked for new pans.

Diane Bowling dropped out of Manzano High School when she got pregnant in the tenth grade. After working at the Albuquerque Ampex plant from 1972 to 1975, she began at Lenkurt in Department 341, where circuit boards were "stuffed" with diodes, resistors, and other components whose leads were soldered to the boards all at once by manual or conveyor-belt immersion in tanks of flux and then melted solder. Bowling added bars to the vats of solder in the "wave solder" machine and reached into the vats of cleaning solvent to retrieve fallen boards.

> I've had my hands in freon many, many times. It gives off a white fog and it turns your skin white. I've had that up to the armpit. I sat with my head over heated freon and solder for hours at a time. One vat was set to 150 degrees, one to 210 degrees, and the third to 100 degrees. There was an even bigger washer for the boards we called "the car wash," and it was about eight feet high. It was a pretty smelly work area. About twice a night I had to climb a lad-

der to get to the top to clear stuck boards. There were several degreasers around the department, also heated. A lot of times they'd malfunction and we'd have freon all over the floor. We had fans when we could find them—about once a week. My throat was irritated constantly, my sinuses were always infected. Beginning in 1978 I had a continuous cold for a year and a half. My doctor sent a note in saying I should stay away from whatever was making me sick, but the company said no restrictions or I'd lose my job, and I couldn't afford to lose the job, so I'd ask my doctor to take the note back.

In addition to the respiratory symptoms, Bowling had odd feelings on her skin. She had the sensation "bugs were crawling on me" or that "a hair was stuck to my face—but when I went to the mirror there was nothing there." GTE's occupational health physician would later react to her reports of such symptoms by referring her to a psychiatrist.

Solvents—Ubiquitous and Toxic

What Bowling and the other workers knew as simply "freon" was any of several mixtures of solvents. One GTE freon contained ethyl alcohol, nitromethane, and 3chloro3fluoromethane. When heated, freons such as this decompose into toxic gases, including phosgene, hydrogen chloride, and chloride. Chronic overexposure to phosgene gas is possible because there are no warning symptoms at low but damaging concentrations. Chronic exposure produces emphysemalike lung conditions; acute exposure causes fluid to form in lungs from five to twelve hours after exposure, producing dizziness, cough, shortness of breath, and feelings of suffocation. The other ubiquitous solvent at the plant was 1,1,1-trichloroethane, or methylchloroform, known to workers only by its trade name abbreviation, "VG." Methylchloroform also produces phosgene gas when heated, even at great dilution. One hundred parts per million of methylchloroform can produce 0.1 ppm of phosgene when heated, which is the permissible exposure limit for phosgene gas.[5] These and many other solvents were labeled only in their fifty-five–gallon drums, but handled by production workers in unmarked containers of smaller sizes. Their toxic effects have long been known.

Because she was born with no sense of smell, Senaida Aragon was designated by her female supervisor to clean out chemical tanks. Such a job assignment contradicts the knowledge, widely shared by toxicologists, that the odor threshold of certain chemicals is important in estimating exposure levels. To smell certain "odorless" chemicals is to be already overexposed (see Marc Lappé's testimony in Chapter 7). Using Senaida in this way took the bell off the cat, eliminating one of the only warning systems the Lenkurt workers had: their noses.

Loretto Herrera, a small man at 5 feet 1 inch and 105 pounds, worked with several other men in the printed circuit lab. The tanks and other PC lab wet area apparatus stood on a slatted wood floor over a concrete pit that sloped like a swimming pool from eighteen inches to five feet below the wood floor. The chemicals were dumped each week from the tanks into the pit, where they mingled and went out a common drain to a waste chemical tank outside. Because he was small enough to get into the tanks, Loretto was the designated PC lab scrubber. The tanks took eight hours to clean, three times a week; the hardest was the nitric acid tank, where a quarter-inch black ring formed, which took much scrubbing. Diego Anaya testifies that several times he and other workers dragged Loretto out into fresh air after they found him slumped to his knees by the tanks, dazed from the fumes. Their requests for fresh-air breaks or respirators were refused by their supervisors and foremen. They were told, "If you don't like it, you can leave."

The PC lab fumes were picked up by exhaust systems that released them through roof vents. Because of the architectural requirements for no visible vents, the exhaust ports were at or below the level of the intake vents for the huge cooling units. Consequently, the coolers recycled the fumes and blew them back into the building's production areas. Dennis Hamilton, who was the assistant plant manager in charge of all mechanical and electrical maintenance from the plant's construction until 1978, says:

> We had a tremendous problem with chemical fog, eating away our equipment in the printed circuit shop. The rectifiers on the west wall . . . [above] where gold, copper, tin-lead, and stannous fluoride plating went on, were constantly being eaten up. I had to replace those monthly. We had to completely resurface the pit below the PC lab floor, because the chemicals would just eat the lime out of the concrete. It would just turn to sand. We'd clean it quarterly and rebuild it yearly. The squirrel-cage fans in the hi-vac units [industrial-size swamp coolers] on the roof would suck fumes from the roof vents and blow it back into the building. Cooled to around 40 degrees, it would sink and form a layer about eight feet thick at floor level where the people worked. We had a tremendous amount of trouble

maintaining the hi-vac units over the plating shop because they were being eaten up as fast as the equipment inside.

Exactly the same sort of roof vent design fault was found to be the cause of serious chemical fume problems at the Signetics semiconductor facility at Sunnyvale, in Silicon Valley, in 1978. Wind conditions and the decorative parapet wall surrounding the roof to hide the vents and exhausts caused vented fumes to form a "toxic swimming pool" up on the roof. At first, Signetics maintained that women workers who complained of nausea, dizziness, confusion, and mood swings were suffering from "assembly line hysteria" or some other personal psychological problem, until male workers started getting similar symptoms. Signetics then had an industrial design firm investigate the fume problem. Three women who were most affected were assigned by Signetics to go around the production area like canaries in a mine shaft. They became known as the "Signetics Three," and after a six-month NIOSH investigation of employee health records found a significant occupationally related health problem, Signetics fired the three workers.[6]

At GTE, management was even less responsive to women workers' complaints. Lila Navarro Banghart worked in one of the closed-off rooms, Crystal Filters, and got plenty of the recycled, fume-laden air.

The vents were right above us. We got backflow from the PC lab through the air conditioning. Your tongue would tingle—it would taste weird—we'd get "instant headache." Some would get nausea and bend over. Vonnie Walker would nearly pass out—I'd have to help her go out on break. The engineers would come, after we'd complain about the backflow. They would claim they couldn't smell anything. We walked out once en masse the first year or two I worked at the plater; we went to the cafeteria and sat there. [A supervisor] told us never to go out again, or we would be terminated. But many times, some of us would have to walk out of there. Pong Sil and the other foreigners stayed, even with her eyes so red. Janet Caudill called OSHA one day on her break. They said they'd have to notify GTE they were coming, so it wouldn't do any good.

The inability of male supervisors to respond to female worker complaints was partially related to Lenkurt's production-at-all-costs management philosophy, but wasn't caused by a lack of complaints from workers. Lila Banghart says, "We complained all the time: 'Dave, this is killing my eyes! This is bad for the eyesight!' But the supervisors just ignored us or insisted it was not harmful." Cecilia Flores Matta had skin and lung reactions to hydrochloric acid, xylene, freon, and inks and binders with silver particles in them.

My rash started with puffy eyes, then spread to my neck, my armpits, and spreads all over if I touch it. I *still* get it [after having left the plant]. I took [chemical] samples to a lab, and my doctor said the materials were causing my rashes. [A supervisor] threatened me, said never to take samples outside the plant. [Another supervisor] told me, "Don't make waves." They scared me—Lenkurt scared me. I made waves, and so I went under and didn't surface again. We needed the money and I had to have that job.

Because of defense contracts, GTE workers had to sign an "Espionage and Censorship" form that prohibited them from "gathering, transmitting or losing defense information." Perhaps their supervisors considered it "transmitting defense information" to get an outside doctor to identify what was causing the women's health problems. Being obliged to sign such a form as a condition of employment, along with another form acquiescing to mandatory overtime whenever management wanted it, demonstrated to new hires that the executives and managers of GTE reserved nearly all prerogatives for themselves.

Stress at Work, Stress at Home

Just as those who worked in the technical, clerical, and managerial offices had separate air conditioning from the production area, they were also separated by class divisions. A common definition of social class holds that upper classes not only have more money than lower classes, but also more power, prestige, and control over their everyday lives. Many of the workers' depositions reveal that their home lives appear to have borne the brunt of the frustration and stress that accompanied being powerless at work. Teresa Garcia and her husband both worked at GTE. She testified to actually liking the solvent fumes because "you could get high for free." She was also aware that there was something ominously stressful about the work:

I used to go home from Lenkurt every day like a raging mad person . . . one minute I would be happy, the next minute I would be screaming, the next minute I would be calm. [*Was your husband abusing you?*] *I* was the one who was abusive. I would get him so angry to the point he would have to hit me because I would lose control all the sudden [*sic*].

Mary Sena also had mood swings:

My boys pointed out to me after I started working at Lenkurt that all of a sudden I'll flare up, just for nothing. Just for nothing that should irritate me, I would, like flare out, like emotion, like stress, like an explosion, or like an outrage, you just want to break things, you just want to hit things. It's uncontrollable. It's like a force in you that you can't control.

Marta Rojas, one of the Silicon Valley "Signetics Three," experienced the same sort of out-of-control, "possessed" states when trichloroethylene and xylene entered her workspace air supply.

These odd feelings were embarrassing, threatening, and hard to understand. Husbands and doctors seemed to doubt that things actually happened the way the workers remembered them. Questioned later by lawyers and medical professionals about these experiences, some workers found medical jargon another layer of confusion. GTE's defense attorney, Carlos Martinez, once asked a plaintiff about her "mood swings." "Mood swings . . . well, they would hit me and make me sort of swing over to the side, or over to the other side, like I was dizzy," she answered him. This and similar misunderstandings, Martinez told me, made him think that some plaintiffs were only repeating lists of symptoms someone had taught them.

The plaintiffs themselves sometimes laughed at the helplessness of some of their feelings at the plant. A recurrent story was of people getting woozy on fumes but trying not to show it. Ellen Kayser told stories that made it difficult for listeners to know whether to feel angry or to laugh: "Betty Campbell passed plumb out, fell off her chair, working with VG [solvent]. Her husband fell off a ladder one day. Sometimes it looked like Alcoholics Anonymous in there. Netty was washing boards in VG, and it looked sort of comical. She hit the floor and jumped back up and looked around."

Health Complaints and the
Deaths of Friends

Despite liking their jobs and the money they made, many GTE Lenkurt production workers began having illness symptoms they had never had before, usually within a year of beginning work. The women in the following six case studies reflect the age, ethnicity, marital and family status, and tenure of employment of the group that would later become plaintiffs against GTE: they began work at the plant near the year 1973 at an average age of thirty-two, and worked an average of nine and a half years in production and assembly. They are among the more articulate and assertive of that group, but their health complaints are not more serious than most who became plaintiffs.

Sadie Miera

Sadie Miera came to work at Lenkurt in May 1973 at the age of thirty-one. She was married and had children ten, nine, and three years old. It was her first and only job. She began in Department 320, Component Assembly, and stayed there ten years. Her first task was "potting," filling transformers with a sealer. She poured the potting compound through paper funnels, varying the flow by putting a pencil in the hole, on trays holding 35 transformers at a time, 4,000 transformers per day. The term *potting* may have sounded vaguely domestic, like washing pots or potting plants, but what she was handling every day was a combination of hydrocarbons, polyolefins, styrene-ethylene, and butylene styrene, a brew of epoxy resins and several types of solvents.

She got rashes and white spots from the potting resin during her first year: "We girls used to say we'd get immune to the potting rash." Some of the spotting was permanent. For the first two years she worked at Lenkurt, Sadie also did all the housework at home, either before or after work, the "double shift" familiar to all employed women. Actually, Sadie had a triple shift, because at work, there was more housework to do:

> Every time they delivered my drum of hardener there would be spills. The handle was always broken, so I used a piece of pipe and a funnel sort of like the kind you pour gasoline with. It got sticky all over the floor. I couldn't stand on it, and the fumes from the hardener were horrible. I asked for the maintenance people to clean it, but my supervisor and foreman said to clean it myself.

From then on, she put cardboard down on her floor and changed it every other week. Her shift mates recall seeing Sadie often on her knees at her work station, cleaning resin hardener with a red can of alcohol, itself a form of solvent known to cause dizziness, irritability, and narcosis when inhaled. The vents over her work tables were always clogged with sticky resin and dust and hair, slowing the air flow to nothing. "The foreman said to clean that ourselves, but there was no vacuum cleaner. So I brought mine from home."

Three years after the plant opened, management changed

the policy on worker assignment. Instead of emphasizing maximum worker and management flexibility by moving all workers from task to task whenever rush orders needed filling, management began to identify good workers and keep them at particular tasks. The underlying reason for both policies, which will be illustrated by Sadie's case, was production efficiency, not worker benefit.

After three years of handling epoxy-solvent combinations, Sadie was chronically sick. She suffered constant nasal and sinus congestion, headache, dizziness, dry cough, sores in nostrils, wheezing, shortness of breath, loss of taste, and nausea. She began having heavy menstrual periods every fourteen days, with considerable discomfort, but felt pressure not to take sick leave or complain: "The reason I never took off is I was the only one who could do that lousy job." When she finally got a helper, the woman got sick and told her doctor she worked with resin hardener; "her doctor asked Lenkurt for a sample of the chemical, and when they refused, I figured there was no use mentioning it to my doctors."

After another two years, Sadie had an enigmatic brush with an engineer from the front office, George Kerr, and her foreman, Jack Wiggins. One day the two men appeared together, watching her work, and asked her, "Don't you think you need a change? Your limit in here should be three to five years." Sadie did not know what to reply; she did not know what they meant by "limit." Kerr offered to help her find another job in the plant. Sadie said in her deposition, "I didn't know if they were joking with me or wanted somebody else in there. If I took a transfer, they could say I refused my other job, and if I refuse a job, I'll be thrown out the door." Wiggins and Kerr made no attempt to explain what they meant or to discover how Sadie felt.

Sadie Miera pressed on with her triple shift, cleaning up at home and at work. Although she had a family doctor she saw between 1973 and 1984, her various problems caused her to be referred to six others before she left GTE in 1984.

Mary Lou C de Baca

Mary Lou C de Baca started at Lenkurt in 1972 at the age of thirty-three. She was married, with children twelve and fourteen, and had worked for about four years as a waitress and a sales and billing clerk. Her first two years were spent in Department 345, Crystal Filters, and the next nine in Department 320, Component Assembly. After she had been at Lenkurt seven months, she began having heavy menstrual bleeding and wearing a sanitary pad every day all month long. In 1974 she had a hysterectomy.

In Department 320, Mary Lou sat across from Sadie Miera, near a tank of heated Chlor-Trimetron solvent. She experienced several of the same symptoms Sadie had, though at different times. She complained to supervisors that solvents ate through the rubber finger cots (fingertip covers) she wore. They issued her more finger cots. In 1978, her work accounting procedure changed to an incentive system, with a required quota and bonuses for production beyond that total. Her health got worse. "I was embarrassed to go to the dispensary every time, because they would say, 'You again?'"

Her husband, a painter, developed a drinking problem and entered Alcoholics Anonymous; a son and his wife had a retarded baby; Mary Lou tried to support them all emotionally while keeping her house spotless. The GTE incentive system offered more money, but more stress as well:

You had to sit there with your neck like this, and I mean you had to work. If you got out of your seat you'd have to rush to the bathroom, back again, sit back down, and I mean with your head like this, and I mean, just eight, ten hours straight, if you wanted to . . . get any money. [We were] always prodded to make more, because if you made a little bit of money that week, it seemed like right away they would bring somebody else to [adjust] the rates, because you were making money and they wanted you to go even faster. . . . I thrived on good attendance. I had only four tardies in fourteen years. . . . [*Did your machinery ever break down?*] No, because . . . I mean, I was the machine, you know? I mean, when I broke down, was when I went to the doctor.

Throughout the 1970s Mary Lou repeatedly went to her doctor, Arturo Garcia, for sore throats, sinus infections, earaches, fevers, gastrointestinal problems, chronic headaches, chronic coughs, rundown feelings, and bleeding gums. In 1975 she was referred to a specialist for thyroid problems, including the constant colds and flus, weight gain and loss, sleepiness, nervousness, hypertension, high blood pressure, mood swings, loss of memory, decline in cognitive ability, and paresthesia (numbness and tingling). Finally, in 1980, she reluctantly accepted a physician's suggestion that she needed psychological help, because he could find no reason for her symptoms. She was given a prescription for Librium, which she used off and on. She had been repeatedly told by supervisors that the chemicals were known to be harmless. Her doctors did not inquire about her working environment. She was examined by fourteen doctors of various specialties between 1976 and 1984.

Mercy Chavez

Mercy Chavez started work at Lenkurt two months after the plant opened. She had been married five years and had children one and three years old. She had four years' experience working at a medical lab in Oregon. Assigned to Department 320, she was loaned out to every production department in the plant except Crystal Filters. Working at a heated freon tank, she began getting nosebleeds, breathing problems, sore throats, and constant mucous drainage. She had septum surgery in 1974 and again in 1976 to relieve the congestion and nosebleeds.

Mercy later worked near Mary Lou and Sadie, mixing the Lencast epoxy and gluing components. In 1974, "there was always fumes that you could see when you walked inside the plant . . . a gray, smoky mist, you know . . . like when a lot of people were smoking." Employees were allowed to smoke on the job in many areas until after 1980. The Lencast epoxy added fumes to that haze:

> I would say it was strong. It wasn't so bad when we had the little fans going, because [they] would pull it out directly away from our face, and the Lencast, once it started to heat, that smelled like a sulfur. Before then, it didn't have an odor, but once it started to warm up, then it would throw this smoke and it would smell. We would all complain, and then like all humans we would come in early and steal fans from people that had them. Little fans, eight inches, maybe ten inches across. Maybe there were ten in the whole department. And at night when we got ready to leave, we would hide them, you know, because people who could get them would lock them up for tomorrow. That's terrible.

Mercy worked over a heated freon tank and was issued a rubber apron and face shield. She complained of the rash that traveled up her arms, and after several months GTE supplied a barrier cream to protect the skin from the solvent and retard solvent absorption through the skin. The cream was kept in a supply area and given only to those who asked for it. GTE had done a time and motion study that was later produced to Josephine Rohr. The study timed each step of applying the barrier cream, in tenths of a second: unscrewing the lid of the jar or tube, applying cream, spreading cream, replacing the lid, putting gloves on over the cream, repeating the process after lunch. GTE's analysis found that the use of a barrier cream took fifteen minutes per day from production—an hour a week, five hours a month, sixty hours a year. Therefore, it was not cost effective.

In October 1980, while working with Lencast epoxy glue, Mercy developed severe blistering of her eyelids, nose, and mouth. Her gums swelled and bled. Her friend Ellen Kayser testified that Mercy looked as if she had measles or skin poisoning; Lucille Guernsey, who had the same reaction, said, "We looked like something out of 'Weird Theater.'" The company physician at Lovelace Medical Center verified her allergic reaction to Lencast, using patch tests; GTE put her on medical leave and began paying her workers' compensation weekly. She asked to be given any other job in the plant, but the company interpreted union rules as prohibiting reassignment unless there was another position of her grade open, even under health criteria.

Management settled her case in May 1982, after she had been on unpaid medical leave for eighteen months, with a lump-sum payment of $20,000 and two years of medical coverage. Lenkurt's workers' compensation insurer, Kemper Insurance, supplied her with a lawyer, whom she met for the first time in the corridor outside the courtroom. She never caught his name. Like many lump-sum settlements, it included clauses releasing GTE from any future liability and declaring that Mercy considered the payment just and fair compensation. Mercy thought it was limited specifically to her lost work time, said yes to all questions, and accepted the settlement, which included $100 for the lawyer who represented her at the hearing. Aside from the Lencast episode, Mercy had consulted nine other physicians over her years at GTE.

Janet M. Caudill

Janet M. Caudill went to work for GTE at the age of twenty-eight in 1974. She had two children, aged three and eight. According to her medical records and her own recollection, she had not had a single health problem prior to her employment at the plant. Caudill worked in Crystal Filters, making little devices tuned to resonate at a certain frequency. She used a lot of trichloroethylene, freon, and isopropyl alcohol at her work station, and sat near a boiling freon tank. She and her friends often had "rotten-egg burps" from upset stomachs.

> They kept upping our quotas. There was this fantastic and constant pressure to produce. We had to get creative to handle the quotas—when we were doing okay, we hid extra crystals behind the machine that we could get and put in our quotas on bad days when the pressure got too high.

A year after she started work at Lenkurt, Caudill hemor-
rhaged for eight to ten days; she got anemia from the blood
loss. She saw a gynecologist, who could find no reason for
the bleeding and took her uterus out. He told her afterward,
"It was too bad we had to do that, because your uterus was
pink and perfectly healthy. But the heavy bleeding gave us no
choice." Janet says she "felt bad for all the women who wanted
babies and couldn't have them, and there was my good uterus
being thrown away." Solvents had been known since the early
1960s to affect the endocrine system, which in women regu-
lates the hormonal functions, including the menstrual cycle,
but no Albuquerque doctor made the connection between
disturbed menstruation and toxic chemicals.

There were about thirty-five workers on Janet's shift, and
several other women in their twenties and thirties had hysterectomies after beginning work there. As the women who had
had the operation found out about the others, they joked,
"We all get spayed in this department." After two more years,
Janet's sinus problems were constant, and she developed high
blood pressure. She had to take medication for a heart arrhythmia, another effect of solvent overexposure reported in
the literature of toxicology. "I had just tolerated the job because we were buying a house and needed the money," Caudill
said. "Then, for a while, I thought I'd just die in that job. But
one day, I felt so bad I just quit and went home. I'd had it
with feeling so bad." Caudill was in the minority of GTE production workers who could afford to quit and go home. Most
lived from paycheck to paycheck, a form of economic power-
lessness enforced by management prerogatives to "adjust the
work force" at will.

Over the four years that passed after she quit Lenkurt and
before the lawsuits became publicized in 1984, Caudill began

to see former co-workers at church and at the shopping malls. She began to realize that a lot of them had had serious illnesses. One of them became a plaintiff, and convinced Caudill to go see Josephine Rohr. Before she did, Caudill sat down and made a list of co-workers on the day shift and their health problems. She was shocked to see the total—eleven women out of the approximately thirty-five who had worked there as long as she did, had had hysterectomies. There were miscarriages, thyroid tumors, nasal operations, and blackout spells sprinkled around the group. National trends in performing unneeded hysterectomies have given women in the United States a 50 percent chance of losing reproductive organs by the age of sixty-five. But Janet Caudill's co-workers had a 30 percent rate, by her count, before they were forty.

Ellen Kayser

Ellen Kayser was one of the few who came to the new GTE plant with previous electronics experience. She was also one of the older workers, starting at GTE in November 1972 at age forty-seven. Her mother was full-blooded Mescalero Apache, her father Scotch-Irish. Her children were grown by the time she started work at the plant. Before coming to Albuquerque, she had toured southern California as the singer for a country and western band, and she still has an outgoing, country talkativeness.

After some brief electronics training at a local Skills Center, Kayser went to work at the new Singer-Friden plant, helping uncrate the tables, shelves, and machinery and set it all up. Managers immediately spotted her skill in solving mechanical problems on the production lines, and made her a head troubleshooter and inspector. Her ideas broke a bottleneck on the line that made heat sinks used in large adding machines, upping production from 25 to 229 heat sinks a day. Her pride in excellent work clashed with a supervisor who asked her to let defective circuit boards be sent out of her department. After taking time off to recover from a heart attack, she decided not to return to Singer and applied at Lenkurt.

A GTE supervisor noticed Kayser's speed and skill in soldering diodes during the mandatory beginner's electronics course at Lenkurt and asked for her in Department 341. There Kayser made two-inch hybrid circuit boards and installed chip capacitors, inductors, integrated circuits, thermistors, and other components on the boards. Ellen excelled at this detail work, but when borrowed for two weeks by the plastics department, where incentive pay seemed easier to get, she asked to remain there even though it was much rougher work:

> It was hard, dirty, dangerous work but it kept me good and busy. And I'm a workaholic, I guess, I enjoy working. . . . We made boxes, spools, tubs, injection moldings, compression moldings, connectors, blocks, face plates, shields, and panels and all sorts of things. It was very interesting work to me, and I did enjoy working in there.

Moving a heavy upright tool cabinet, with drawers of tools open on the side away from her vision, Ellen broke her wrist: "And the box tipped over, and that handle on there caught my wrist where I couldn't turn loose, and it just twisted on around." She had the GTE nurse put an Ace bandage on it and continued operating power tools for three days. "It finally swolled so bad they sent me to the doctor. It hurt like the mischief, but I and the nurse didn't realize it was broken." She also cut off a finger on a table router:

> And this I blamed on the engineer that had us do this job, and I told him at the time, I says, "Kelly, someone is going to get hurt seriously with that, because you have no safety device to hold that part." They didn't fix it, and it was me that got it, and I had been warning him that that's what would happen. And as you can see, it's way shorter than the other one because of the piece that was buggered up right bad in the joint. And I do have a pin in it yet from that.

In 1976 Kayser was transferred to Department 320. She sat near barrels with disturbing labels:

> It says, "Can Cause Death." I mean, just that simple, "Can Cause Death." I worked as far as from you to me from those barrels all the time. [*What was in the barrels?*] I don't know, but when it runs over the top of the tank, it peels the paint right off the tank.

In 1979, Kayser became "deathly ill":

> I kept losing my voice and losing my voice, and at first it was periodically. The nurse referred to it as laryngitis, but I had no sign of cold or anything that remotely resembled a cold . . . and the last time it happened was on a Friday, we were heading back to the plant from the bank . . . and she and I were talking, and right in the middle of the conversation, I mean, my words just shut off, I couldn't get a

sound out. And she said, "Well, finish telling me." And I
couldn't . . . and it wasn't five minutes later that this pain
hit my throat, if you can feature in your mind or imagi-
nation what it would feel like to have a giant claw grab
ahold of your throat and rip it out . . . it was excruciating.
I couldn't even catch my breath to scream. I couldn't have
screamed if I'd wanted to, I had no voice.

The illness was thyroid cancer, for which she had surgery
and radiation treatment that year. Divorced, with no alimony
or insurance, she returned to work almost immediately, unable
to afford to take a longer recovery without income.

I was out exactly a month with that [surgery]. I had to go
along there [at work] holding on to things to walk because
of the weakened condition I was in. A size ten pant was
big and baggy on me, if you can feature that in your mind.
But I still continued working. . . . The following February 29th, I started again getting so deathly sick, and I had
this feeling like I was plugged into the wall and somebody
had pulled the plug. I just felt like if I was to relax for one
instant that I would just literally melt into the floor. I didn't
feel like there was a bone in my whole structure, and the
hurt was terrible, all in my upper chest and in my throat,
the side of my neck, and in behind what I guess is the mas-
toid there . . . and it seemed like it just radiated back in
behind this eye. And that continued—I lost a total of fifty-
eight pounds from February 29th to March 28th . . . and
through it all I have kept going to work. I try not to miss
any at all. I try to do my job to the best of my ability. Like
I said, as long as the Lord loans me breath, I want to work
and earn my way. . . . I fluctuate from a size eighteen pant

to a size ten, back and forth, and it's really miserable, I'll
tell you. . . . And the headaches are always there, and the
dizziness and nausea.

[*Do you think these problems are connected?*]

It set us all to wondering why so many from that general
area started getting so ill. Several people died. One woman
died in my arms at St. Joseph Hospital, Isabel Romero,
she died in my arms. And she died in April, prior to my
going into the hospital in September. . . . And the supervisor that used to be on the line working with Isabel was
Lupe Luna, and she lost a breast to cancer right after I
had the thyroid cancer. And right across the aisle from
us was the maintenance department, and Steve—I can't
think of Steve's name, bless his soul. But Steve died of
cancer. Jim Kelly in the tool crib right directly across from
where I was working has lost a leg, or maybe his life by
now, I don't know. Cammie Wright used to work in that
area when it was Department 390. She was a supervisor.
She died. She went into the hospital the same time I went
into the hospital. Dot Tuma worked right alongside of us
in the Lencast and Plastisol and them other things . . . but
Dot is no longer with us. She died of a brain tumor. And
Mary Anaya, she's dead too. And then there was Margie,
they removed a brain tumor from her too, but she's still
with us. There is Ruby Shannon, and she's all crippled
up with cancer. There was so many people. There is Ed
Wooten, used to work in 390 with Cammie Wright, and
they removed a tumorous growth from his shoulder. And
there has been several people that worked with chemicals
in the PC lab, the wet area, nine out of ten of these people
have had problems of some kind. And I cannot speak for
them, and I'm not trying to, but you asked how did I con-

nect all of this. . . . Irene Baddy, she used to be in charge
of moving those chemicals, those barrels and drums with
a forklift. And I walked in the bathroom one time and this
lady was frantically washing her hands and shaking like a
leaf. And I said, "My gosh, Irene, what's wrong?" And she
couldn't hardly get any words out. I put my arms around
her and calmed her down, and she said, "They're making
me change all the labels on all those barrels out there because OSHA is coming to check." And I says, "Well, Irene,
surely they won't make you do anything that would harm
you." And she says, "Yes, but some of that spilled on me."
That woman couldn't calm down and was removed from
the plant, taken to a mental institute. The last I heard
of her they had taken her to some sanitarium or something. She wasn't exactly the smartest person in the world,
but she knew that something was radically wrong, or why
would they have her change the labels?

Ellen Kayser was devastated by the death of Isabel Romero,
a quiet Pueblo woman. Though Apache and Pueblo peoples
were historically enemies, there was special closeness between
the Native Americans of any tribes at the factory.

One day at work she was looking as gray as death, and I
said, "Isabel, my Lord, are you all right?" She said she had
cancer and would only be able to work a few more days.
She went out to her mother's at Santa Ana Pueblo when
she couldn't work any more. Then, one day when they had
brought her in to town to the hospital, I went to visit her
there, it was a Wednesday. She asked me to sit her up. She
looked at the clock, and said, "One more day." I looked at
the clock, and it said five 'til six.

Next day at work, I asked another Indian friend, Inez, to come with me to see Isabel again right after work. We hurried over there—it was nearly six when we got there. Isabel was rational 'til the last minute. She said, "Oh, Ellen, you're here." I was standing beside the bed; she asked me to sit her up. I took ahold of her one hand, and her sister Lucy took the other. She looked out the window; it was clear as could be. She said, "The clouds will come to take me." Then she said, "I'm so tired." I asked the Lord to take her right then. [Ellen is weeping, her strong voice choked into falsetto.] She was looking straight up at the ceiling. I took and closed her eyes. And we laid her back down. When her mother realized what had happened, her mom grabbed me and clung to me and said, "You're my daughter now." That was April 26, 1979.

Federal Investigations, Riots, and Strikes

In 1978, near the end of the first International Brotherhood of Electrical Workers (IBEW) contract, production worker frustrations boiled over in the most dramatic labor strife in New Mexico since the Silver City mine strikes reenacted in the 1955 film *Salt of the Earth*.

A supervisor, Audie Kaufman, claimed four of his workers —a man and three women, one of whom was the shop union steward—followed him and harassed him as he left the plant after their shift was over on Thursday, May 5. The nature of their dispute has never been made public. The next day Kaufman informed the four at the beginning of the shift that he had placed written reprimands in their files for harassing him. The four workers felt threatened—under the union contract, written reprimands could be grounds for dismissal. By coffee-break time, the four were angrily discussing the file memos with co-workers in their area; they reached a consensus that they must demand the removal of the reprimands. Kaufman refused, so the workers walked off their posts and went into the front office to demand the removal of the memos. There was shouting back and forth.

When management refused action, the group returned to the production area and gathered in the cafeteria, where workers from other departments, taking coffee breaks, heard of the conflict and joined them. Resentments boiled over as people stood to make speeches. Work was effectively stopped throughout the production area. Several worker representatives, including Juanita Larkin, president of the IBEW local, met with management in the front office to negotiate. Over an hour passed; each side refused to budge from their positions on removal of the reprimands.

The group in the cafeteria swelled to about two hundred and became boisterous, chanting about human rights. Though workers neither damaged property, nor threatened or made physical contact with management personnel or nonparticipating workers, Larkin says, "It was pretty much a riot." With no warning, Lenkurt management then called city police, and at 11:47 A.M., the SWAT team appeared in full riot gear with dogs, shotguns, and batons and ordered the demonstrators to vacate the building. Larkin says,

> We were shocked. The dogs were barking and snarling, and they took them all over the plant, clear back to the wet area, looking for people hiding. They yelled at us to vacate the building, and as we filed out, people from the front office videotaped us so they knew who participated. When the swing and graveyard shifts reported for work later in the day, they found the building locked, with police cars at the gate.

The day of the demonstration, GTE suspended 350 workers for illegal work stoppage, via telegrams from San Carlos. The next day workers demonstrated and marched with picket signs outside the locked plant, and Lenkurt executives again called city police. The mayor, David Rusk, son of former Secretary of State Dean Rusk, was approached by his deputy police chief, who complained that he felt Lenkurt was trying to use his riot squads as plant security forces and strikebreakers and he was fed up with it. Rusk and the deputy chief met at the plant with GTE management to inform them of the city's unwillingness to be so used. Tom Wortman, plant manager, raged at Rusk, "You'll never get another plant to move to Albuquerque after this!"

On May 10, GTE claimed to have gotten a court order directing union workers to return to work, asking $360,000 per day from the union in actual damages and $100,000 per day in punitive damages. No record of such an order can be found in NLRB or local federal court files, but the story was detailed in the press. Juanita Larkin ordered her members back to work, ridiculing the company's request for $460,000 per day in penalties: "We haven't even got $20,000." But Lenkurt officials kept the plant locked, claiming suspended workers could not return to work. The workers appealed to the NLRB, which sent a negotiator. By May 21, an agreement was reached for all suspended workers to return to work, losing pay for the walkout/lockout period, but with all associated reprimands removed from their files. According to the agreement, twenty-four workers would be fired.

In October, Thomas Blinn, general manager of the plant, announced he had found the reasons for the worker unrest, and it turned out to be the same charge leveled at those portrayed in *Salt of the Earth* twenty-four years earlier: "We have a Lenkurt Communist Party," which he charged would damage Lenkurt's reputation and Albuquerque's ability to attract new industries. One week later, the *Albuquerque Journal* reported having received a two-page press release from the "Mao Tse Tung Memorial Committee and Revolutionary Communist Party USA," urging sympathizers to attend meetings in New York and San Francisco at which organizers would "denounce the current Chinese leadership." The newspaper asked Lenkurt management and the union if the paper could review any Communist material in their files, to compare it to the press release. Union spokespeople and workers angrily denied that a Communist conspiracy was responsible for their demonstrations, and pointed to the conditions at the plant leading to such extreme dissatisfaction, including an unworkable grievance procedure. Management also refused to open its files to the press. Larkin says there were between ten and twenty workers calling themselves Communists, and they had little luck, in her experience, converting any of the other production workers.

Three months later, on February 1, 1979, while Ellen Kayser lay recuperating from thyroid surgery, 1,450 of the 1,900 employees at the plant—the entire unionized production work force—went out on strike, demanding that the new union contract, about to be signed, contain written confirmation of new grievance procedures. Two weeks later, the union signed a new three-year contract, which stipulated a 36-cent raise, some cost-of-living steps, a new dental plan, and one additional annual holiday.

OSHA Inspections at the Plant

The IBEW union filed several grievances with management over health and safety issues during the 1970s and called in federal OSHA inspectors. Elizabeth Marilott, secretary to the local OSHA director in the early 1970s, recalls that "fifteen gals came into the office in a kind of mass protest in 1973 or '74. They complained that they were working with dangerous chemicals." OSHA's inspection history for GTE, obtained through the Freedom of Information Act, lists two inspections in 1974, one each in 1975, 1976, and 1977, three in 1978, and four more in 1979, indicating a steep increase in worker complaints in the year of the demonstrations. According to the OSHA documents, two citations were issued in 1976, and an initial penalty of $800 was dismissed; three citations were issued on one visit in 1979, and an initial penalty of $200 was also dismissed. Federal OSHA issued no other citations for unsafe practices.

Industrialists and small entrepreneurs decried all federal regulation during the 1980s, but OSHA's record of setting standards for workplace hazards and punishing violators should cause no manager to lose sleep. In its first sixteen years, as transnational high-tech corporations assumed a larger and larger share of the economy, OSHA issued only eighteen health and twenty-three safety rules, which its director in 1987, then Assistant Secretary of Labor John Pendergrass, called "embarrassing."[1] Enforcement was even weaker. The Justice Department brought to trial only one of the twenty-four cases OSHA prosecuted under the Reagan administrations. No one has ever gone to jail for violating workplace safety standards in the United States.

A Caring Labor Department Investigator

Two federal investigations began in 1979 that had some impact on GTE Lenkurt production workers. Lee Leyba, who had just retired from the Coast Guard, took a job with the local office of the Office of Federal Contract Compliance (OFCCP), a branch of the U.S. Department of Labor. One of Leyba's first cases was a complaint from Connie Ruiz, a GTE production worker with five years' experience who was denied promotion from assembler to technician even though she passed the company test. Ruiz decided to complain to the U.S. Department of Labor. Her complaint notes that there were 168 male technicians and testers at GTE at the time, and no females. The Lenkurt test required a score of 120; Ruiz made 105 the first time, 110 the second, then 180 the third time, which should have qualified her not only for technician but tester as well. However, her promotion was denied and Lenkurt hired males from new applications.

After interviewing Ruiz and visiting the plant to see her employment file, Leyba was struck by the number of rashes, episodes of nausea and gastritis, high blood pressure, and other complaints he found in worker medical files. He wrote a memo to OSHA, also within the Labor Department, urging that the health conditions at the plant be investigated. Leyba soon filed OFCCP reports about several other Lenkurt workers who appeared to have been fired because they became too ill on the job to continue working. Leyba says that men whom he would rather not identify tried to intimidate him. One told him that Leyba "didn't know what kind of street fight he was getting into" by pushing such cases against GTE. Leyba replied, "If a fight is what you're offering, I grew up in the barrio. I can handle myself. If you want, I'll go down to any street or alley with you right now." The men did not press the issue.

GTE Lenkurt has blamed union-sponsored rules for the dismissals Leyba investigated. The union fought to curb management's ability to transfer workers arbitrarily, and negotiated a contract provision allowing a worker to be transferred only if he or she bid on an open position in another department. The OFCCP cases involve several workers who had allergic reactions to Lencast epoxy and other chemicals; GTE management claimed they could not transfer them if there was no open slot and thus no formal bid to transfer. According to Judy Mathes, an attorney in the civil rights division of the solicitor's office in the Labor Department in Washington, D.C., local OFCCP offices are required to conciliate such disputes unless they are too big or nationally significant, in which case they are referred to Washington for settlement negotiations. After sitting in the Dallas regional OFCCP offices for several years, some twenty cases of illegal firings did find their way to the solicitor's office.

An OSHA Inspector Takes the Revolving Door

The most significant consequence of OSHA visits to GTE occurred in August 1979, when senior inspector Jack Lacy went to work for Lenkurt instead of reporting the many violations he discovered there. Lacy was sent to the plant to inspect the printed circuit lab's dispensing and storage area, where Loretto Herrera so often became faint earlier in the decade. In his inspection report, obtained via the Freedom of Information Act, Lacy wrote that the company had ordered a new exhaust hood and no further surveillance was necessary, noting that followup responsibility was assigned to the inspector, Jack Lacy. One month and four days later, Jack Lacy began work as the new safety and loss control officer at GTE Lenkurt.

According to a recorded interview conducted with Lacy in 1985 by plaintiffs' attorneys and transcribed by a court reporting service, Lacy says that he found 150 serious violations and 300 minor ones when he did a walk-through inspection of the plant on the July 1979 inspection visit prompted by

Leyba's memo. Instead of informing his OSHA superiors, and thus triggering an unannounced plantwide inspection, Lacy told GTE human resources manager Fritz Hannah about his findings, in confidence:

> I told Fritz if OSHA walked in here and found these viola-
> tions, you'd have $65,000 to $75,000 in repairs to do, plus
> $180,000 in fines for the 150 major violations, and another
> $30,000 to $40,000 in fines for the 300 minor violations.
> He said, "You want to quit the federal government?" I said
> I'd been wanting to for over four years, had already lost a
> wife over it, so why not? So he offered me a job.

Given OSHA's weak record of enforcement, perhaps it was better for GTE's employees that Lacy went to work at the plant. According to the Kansas interview and to his official deposition and the depositions of Fritz Hannah and several plaintiffs, Lacy proposed and administered several major safety and health reforms upon taking the job at Lenkurt. He reduced Department 320, the Component Assembly operation where so many worked who subsequently became plaintiffs, from seventy-five to fifteen people; ordered new exhaust equipment for several departments; gained veto power over new plant modifications; set up screening programs for high blood pressure, eye infections, and diabetes; and initiated the medical support program with Lovelace Medical Center, replacing Lenkurt's company nurses with Lovelace physician's assistants, supported by weekly visits by a physician. The physician's assistants were instructed to make appointments at Lovelace for workers with serious problems.

The recordkeeping was not foolproof, as Betty Rogers's story illustrates. Rogers experienced one of the most notorious collapses at the plant, turning blue with an irregular heartbeat. Heart arrhythmia is known to be an effect of over-exposure to organic solvents. She was taken by ambulance to Lovelace; she was given three blood transfusions, one at GTE, one on the way, and one on arrival. During the trip her blood pressure measured 80 over 0. After three hours of ob-servation, she was sent back to work. Neither GTE nor Love-lace could find records of this incident, causing Rogers to tell Josephine Rohr, "Maybe I am crazy. This probably didn't hap-pen." Urged by Josephine to search her papers, she found the ambulance bill. GTE and Lovelace then found their records when presented with Rogers's own record.

In 1980, Jack Lacy had one of the physician's assistants compile an informal epidemiological study of worker health complaints. For the month of October, he found that the dis-pensary had 147 visits for rashes, 57 visits for hypertension, 48 asthma complaints, and 12 diabetes problems. "When I showed our list to [a Lenkurt executive], he said, 'Don't put any of this stuff in writing! I've never seen that list, under-stand?'"

In his interview, Lacy verified all the problems with chemi-cal exposure that workers had been complaining about and later reported in their depositions. Renovating and repairing exhaust systems, he found chemical residues in the form of thick liquor, some dripping like syrup, some hardened like lacquer: "I think it was mostly aldehydes, heavy metals, bad stuff. I drew up a P.O. [purchase order] for about $700 to pay a lab to see what it was, but they said it was too expensive and wouldn't let me test it."

Lacy contacted the loss-control survey organization used (and owned) by Kemper, the plant's insurer. National Loss Control (NATLSCO) had inspected the plant at least twice yearly since its opening. NATLSCO's reports had until then emphasized kinetic hazards like motors, ladders, and wet spots on the floor. Lacy found one inspector for NATLSCO who knew some chemical toxicology and asked for help in analyzing some of the fume problems workers complained about. The two men found that wires insulated with polyurethane, when dipped into boiling solder above the insulation, gave off hydrogen cyanide gas. Lacy had the process altered. In the Kansas interview, Lacy said that he had considered the health hazard of the ventilation system in the epoxy area, where Sadie Miera worked, to be 10 on a scale of 10, with that in the crystal filter shop between 8 and 9. He says management declined to fix the epoxy area ventilation—they decided to wait until OSHA caught it.

Lacy confirmed plant maintenance engineer Dennis Hamilton's description of the way the building's ventilation system recycled the fumes. Lacy found that under prevailing westerly winds, there appeared to be little problem, but when the wind shifted to the east, fumes from the chip capacitor shop invaded the rest of the building, and when winds came from the north, fumes from the crystal filter shop were blown back in. He asked for a wind sock, was refused, and installed his own small weather station, which soon broke; "so I had to watch the American flag out front to check the wind."

Lacy also found people warming their lunches in the ovens used for curing photoresist inks on circuit boards. In the Thick Film lab, "one pot had 67 different inks, containing manganese, silver, gold, terpene solids, who knows? I had MSDSs [Material Safety Data Sheets] for only twenty to twenty-five; the room was not as big as my coat closet, and it was worth $3 million." Lacy claims that in investigating and remedying these violations of health standards, he had the support of some top-level Lenkurt executives in Albuquerque but implacable resistance from others.

According to his interview, Lacy tried to research and to manipulate company insurance policies to have positive effects on worker health. He found that Lenkurt was self-insured through two insurance companies: Traveler's handled the nonoccupational health complaints and Kemper the compensable occupational ones. Lacy claims to have channeled some compensable work-related injuries to the Traveler's policy to get several workers coverage when they were beyond the statute of limitations for the Kemper workers' compensation policy. He suspects others in the past had been routed through Traveler's to avoid having to report them to OSHA as occupation-related. Lacy wanted to support his 1980 informal epidemiological study by reviewing the insurance claims paid by the nonoccupational policy over the last few years. Such a review might have shown elevated frequencies of various diseases perhaps traceable to workplace causes. A GTE corporate subsidiary, Johnson and Higgins, was supposed to furnish plants with printouts of their yearly frequency of insurance claims, but Lacy says all his requests for the nonoccupational printouts were ignored.

Lacy says that Lenkurt's former assembly plant in San Carlos was much worse in ventilation and design than the Albuquerque plant, because the operation had been put in existing buildings designed for other uses. "[The] Albuquerque [plant] was an absolute queen compared to San Carlos," he said in his Kansas interview. Lacy says he was sent to San Carlos to dispose of seven outstanding occupational disease claims when manufacturing was closed down there around 1980. "I had blank checks with me, with $150,000 in the account. Orders were to close 'em out, no health rehab money, no long-term medical, nothing." In October 1980, after Lenkurt had laid off 500 of its Albuquerque workers, Tom Blinn, the vice president who had discovered that Communism was behind the worker rebellion two years earlier, announced, "We're trying to remedy a workforce morale problem caused by autocratic management practices. These morale problems were caused by poor communication, and we're starting a plant newsletter and an open-door policy to remedy that."

The end of his health and safety reforms, Lacy claims, came when GTE's central corporate leaders called all plant safety officers to headquarters and warned them that they would all be considered "liabilities" if their safety and health programs were not operating "right"—at lower costs—within eighteen months. Lacy recounts with disgust a subsequent visit to the Albuquerque plant in 1983 by an executive from Automatic Electric, a Chicago-based GTE division that had responsibility for Lenkurt at various times including the period in 1983. The executive came specifically to investigate Lacy's health-monitoring activities. Lacy thought he was a typical Chicago fat cat. He basically told Lacy, if it ain't broke, don't fix it. He said they were not going to make any changes on anything unless they got caught by OSHA, but that if they got caught, they would fix it.

State OSHA's Single Investigation

On October 30, 1980, production worker Maureen Keller reported problems with Lencast epoxy to state officials in Santa Fe. Sam Rogers, then an inspector and now chief of the state OSHA agency, said he never would have sustained the violations and citations without Jack Lacy, whom he knew from when he had been a federal OSHA inspector. Lacy gave Rogers MSDSs, patch tests from workers' medical files, past medical records, and so on. Rogers thought Lacy had gotten on top of things, although he didn't approve of everything Lacy had done.

The state OSHA file on GTE shows the company paid a $1,000 fine for subjecting workers to the epoxy against manufacturers' warnings, and that the company spent $30,000 in modifying exhaust systems, work areas, and material-handling procedures to abate the harmful effects of their quick-drying epoxy.

GTE Heads for the Border

Lacy was laid off in 1984, six weeks, he says, after GTE laid off the safety director of Lenkurt's El Paso plant, Jim Jordan, along with others, leaving GTE's telephone divisions with a health and safety staff composed of a single industrial hygienist based at the Connecticut offices. And so the sum of institutional response to Lenkurt worker complaints in the thirteen years prior to the avalanche of lawsuits beginning in 1984 was $1,000 in fines from twelve OSHA inspections; probably less than $50,000 in modifications urged by Jack Lacy or ordered by state OSHA; a scattering of insurance settlements for acutely affected workers; and the cost of an improved on-site health response program with Lovelace Medical Center.

The company was already putting its Albuquerque health complaints behind it at the time they got rid of Lacy; Lenkurt acquired land and buildings in Juarez in 1982 immediately after Mexico devalued the peso. This created in one fell swoop a pool of impoverished and eager workers who would work for 5 percent of U.S. wages. GTE moved the most menial, and dirty, assembly departments to Juarez during 1983. Unions in Mexico—in particular, unions in the export-processing zone near the U.S. border—are virtually extensions of management, and there are no OSHA and OFCCP inspectors to worry about.[2]

The GTE Lenkurt plant record shows that the OSHA inspections increased sharply at the same time the union made its strongest collective resistance to management. Clearly, attention to worker health came only in the context of attention to workers as an organized body able to stop work at the plant and bring public scrutiny via the press. As isolated individuals, workers were not heard, even strong women like Ellen Kayser, recognized for making contributions to chemical spill prevention. Robert Howard put his finger on it in *Brave New Workplace*: when those in control believe so devoutly that their high-tech world has utopian possibilities, the Sadie Mieras and Ellen Kaysers of the work force may as well have lost their voices permanently.[3]

Josephine Rohr Builds a Case Against GTE

In August 1984, GTE Lenkurt had been moving production operations to Juarez for a year and a half; Jack Lacy and other GTE division safety officers were about a month away from being laid off; and Amy Cordova Romero, after weeks in the hospital, walked into Josephine Rohr's office on Rio Grande Boulevard near Old Town in Albuquerque.

When she began investigating the list of Amy's co-workers with health problems, Josephine Rohr tried to interest her parent firm in the case, and when the plaintiffs began to proliferate, she asked for clerical support. The firm refused and advised her to drop the investigation and get on with more productive casework. Rohr decided to end her relationship with the firm and strike out on her own, with Amy's case and referrals from her two years of work in workers' compensation, immigration, and collections law. She hoped she could build a private practice that would give her the freedom to follow her conscience before her billings evaporated.

Rohr quickly found a network of women activists that proved invaluable. The head of New Mexico OSHA at the time was Carol Oppenheimer, a lawyer who had stepped up OSHA's advocacy of workers' rights when she took over the agency in 1980. Oppenheimer told Rohr to call Lynda Taylor, a veteran antinuclear activist and political organizer with experience in the Seabrook nuclear power plant controversy in

Massachusetts and the Karen Silkwood plutonium contamination case in Oklahoma. Taylor had moved to Albuquerque to do environmental and electoral organizing and had telephone numbers of activists in Silicon Valley, including Mandy Hawes, a lawyer who had represented perhaps one hundred Silicon Valley electronics workers in toxic exposure cases and won about eighty of them. Hawes had been the lawyer for Marta Rojas, Cathy Bauerle, and Kathi Hee—the Signetics Three—a few years earlier. At this time Rohr thought cancer was the main disease caused by the chemicals at GTE Lenkurt. Although most of Hawes's clients had had other complaints, she legitimated Rohr's investigation by assuring her that carcinogenic chemicals were used in microelectronics work.

One of the first names Amy had given Rohr was Mercy Chavez. Mercy immediately came to see Josephine, bringing her medical records. Rohr called GTE to inquire about Amy and Mercy, and was referred to Kemper Insurance Company, whose spokesperson quickly denied that any disease could be caused by anything at the plant. After talking with Hawes, Rohr filed separate claims for Amy and Mercy in state district court in Albuquerque under the state's Occupational Disease and Disablement law, alleging that working with toxic chemicals had disabled both women. A week later, she filed a claim for Yolanda Lozano, and filed four more by November, two months after Amy came in for help with her finances.

At the time she filed the first two cases, Rohr had about sixteen former GTE workers signed to retainers as clients, and eighteen other names of GTE workers known to have cancers or thought to have died from cancer, including two supervisors and an engineer. The plaintiffs spoke bitterly but cautiously about the intolerant management style and smelly chemicals they remembered at GTE. Some were shy and reticent, having never spoken to a lawyer in their lives; others were vociferous. Feelings of grief, loss, and fear began to surface and became the plaintiff group's common mood as they made lists of ill or dead workers and reminisced about them. Fear of a "cancer epidemic" became central for them. Having lost her own son to cancer six years earlier, Josephine shared their feelings of vulnerability and loss.

Ethnicity and gender played a strong role in the plaintiffs' trust of Josephine. Beginning with Amy and Mercy, nearly all the first twenty plaintiffs were Hispanic. They were able to use English or Spanish in telling Rohr about their physical and emotional problems, including their reproductive system dysfunctions, which they were unlikely to have discussed freely with a male. For some, Josephine was the first person they had confided these details to. Rohr found that she was serving as confidant and counselor for people who had few, if any, specialists in their social network.

The GTE assembly workers who joined the case early on came from a stratum in which there were seldom personal physicians, lawyers, insurance or financial advisers, and certainly no psychological or psychiatric counselors. These working women had doubts about psychological therapy, which they thought of as "admitting being crazy." They were particularly concerned that such a label would make it difficult for them to find and keep jobs—and they had to work. Only 2 of the first 115 clients had the economic option of moving in with parents or other family members; all others were self-supporting heads of households or part of struggling two-worker families, and most had made maximum efforts to keep working despite multiple health problems.

Rohr became deeply involved in their stories:

Some of them had never even told their husbands or other relatives some of their problems. One woman had to wear Pampers [because] she worked on the assembly line and was bleeding so much that normal pads wouldn't last through a shift. There are quite a few stories in the depositions of sudden excessive menstrual bleeding, and wearing two and three pads at a time, and so on. Work rules forbade them from getting up or stopping the line. Some of them felt they could not complain for fear of burdening their families, and here they were working like mules for them! They would tell me these personal things and start to cry, and I would find myself crying right along with them.

Investigating or Soliciting?

As the work of filing cases increased, Rohr hired two recent law graduates, Rudy Martin and Juan Gonzales, to follow the chain of referrals given by clients. Not polished investigators, Martin and Gonzales exaggerated the data available to them. On October 4, Rudy Martin called the Centro Familiar health clinic to ask about client Margaret Gonzalez's lung history. In the clinic's very detailed records, a clerk noted in Gonzalez's file that Martin claimed that 20 to 30 GTE workers out of a list of 150 had cancers, "mostly lung." There was no documentation at the time of his call that even suggested that number of lung cancer cases.

GTE's Albuquerque defense attorney, Carlos Martinez, charged that Lenkurt employees reported that Martin and Gonzales were trying to solicit clients—a violation of legal ethics—rather than merely investigate the claims of those

Rohr already had. Rohr was walking a fine line. After Martinez's complaint, she terminated the young men and resumed calling new workers herself. She could not afford the investigators anyway, since the GTE investigation began to take over her routine practice, filling nights and weekends also. Many of the workers she contacted were afraid to meet her openly; she met some in parked cars or at convenience stores and restaurants, sometimes at midnight or 7:00 A.M. when shifts ended.

In October, Josephine received a visit from Jack Lacy, just laid off by GTE. According to Rohr, he acted supportive, intimating that he knew all about the chemical overexposures and sympathized with her clients. Rohr told him generally what she knew so far, and what she was planning to do. Within a week, she received a telephone call from Carlos Martinez: Lacy had appeared at his office, offering to reveal Rohr's legal strategy for, say, $5,000. Martinez said he had refused, thrown Lacy out, and immediately informed Rohr so as to protect the integrity of the proceedings. Nevertheless, within a month Rohr received a letter from Martinez's firm informing her that GTE had entered into a contract with Lacy that limited his right to divulge his knowledge of GTE's proprietary information, including processes, chemicals, and procedures. The letter threatened that GTE would file tort charges against Rohr if she spoke to Lacy again.

Rohr sought help and advice from the state tumor registry, based in Albuquerque at the University of New Mexico Medical Center. One of the few population-based tumor registries in the country, it regularly sends investigators to all medical facilities to gather information on every malignant tumor diagnosed in the state. Dr. Charles Key, the registry director, declined to extend her any services, saying he felt "it was inappropriate to use a state agency to benefit a lawsuit." Soon, however, Key's office cooperated with GTE's defense attorneys and expert medical witnesses, who had several thousand dollars to pledge toward a study.

Rohr Asks the State Epidemiologist's Help

By January 1985, Rohr had filed twenty-two claims and had twenty additional reported cases of serious health conditions. Responding to press coverage of the first twenty suits, GTE asked Dr. Harry Hull, the state epidemiologist, to investigate the plaintiffs' claims of excessive cancer at the plant. Hull asked GTE's defense attorney and Rohr for any information they had on the list of plaintiffs current at that time. Carlos Martinez provided the information he had been supplied by GTE, which did not include the total number of workers over time at the plant, their distribution among the departments, or their complete medical records. Rohr asked Hull if he would guarantee the files' confidentiality if she sent them to him, since she feared turning over her files to anyone. Hull refused. Rohr later explained:

> I was all alone. I needed to keep secret what other problems I was pursuing, since I hadn't filed lawsuits on some people yet. I was realizing that there were other conditions besides cancer, and I was afraid if GTE knew what I was investigating, they would get to those women and intimidate them into refusing to talk to me. Remember, Carlos Martinez had threatened me with "tortious interference" for talking to Jack Lacy. Besides that, I had been followed at night for weeks by a shabby-looking man in a big, older

car. I started carrying a gun. My second husband had been in security. I knew when I was being followed, and what to do about it.

Hull issued his report in the form of a Memorandum to Epidemiology Files, March 15, 1985. The memo said, "This investigation . . . did not indicate an unusual pattern [of cancer]. The assessment, however, was based on a limited data set." The investigation was based on state tumor registry files current to July 1984, thus excluding several plaintiffs whose cancers were diagnosed after that date. No search of tumor registry files was attempted using GTE employment records; the investigation simply examined the records of twenty plaintiffs, found eight verified cancers, and concluded that the "diverse spectrum of cancer types . . . would indicate numerous different causes and point away from a common etiologic agent."

Carol Oppenheimer of state OSHA was one of twelve people listed at the report's end as interested parties getting copies. Oppenheimer wrote a memo to her superior, Neil Weber, head of the environmental improvement division, to put her opposition to the report on the record. She felt the report had been issued too hastily and was based on insufficient data. According to Oppenheimer, Hull was quite upset when he saw Oppenheimer's dissenting memo.

Rohr felt that it was good that the state epidemiologist was investigating the cancers—even if only at the request of the defense—but wanted someone in New Mexico to investigate the plaintiffs' other problems as well, which seemed epidemic. As the state's chief public health officer, Hull seemed the logical person to do such a study, or at least to show some interest and refer her to other specialists. Rohr was operating in an adversarial legal system that bound her to protect her clients' interests. The legal doctrine of attorney-client privilege recognizes the need for restricted exchange of information prior to a win-or-lose fate in court, which is why she had asked Hull if he could examine the files at her office.

A Common Cause for Multiple Symptoms

Hull's report did not dampen Rohr's suspicions of toxic effects, because new plaintiffs kept supplying grotesque symptoms. Most of the following symptoms were in the workers' medical records long before they became plaintiffs. Sylvia Sena had bleeding in the brain stem that caused partial blindness and paralysis, and her Albuquerque doctors had to refer her to a Boston neurosurgeon for "stereoscopic proton beam therapy." The Boston specialist said her tumor was an extremely rare one. Amalia Romero, who had never worked anywhere before GTE, had pus-filled boils in her groin and a kidney destroyed by an infection after working only eight months; Roberta Tena, groin boils the size of oranges; Grace Wessel, cysts the size of cantaloupes; Carlotta Leon, an abdominal tumor the size of a volleyball. These gruesome conditions, in addition to the widespread histories of subjective symptoms such as nausea, malaise, anxiety, numbness, dizziness, headaches, chronic colds and flus, sore throats and sinusitis, eye irritations, central nervous system anomalies, and menstrual dysfunction, did not seem to Rohr and her plaintiffs to be simply the usual processes of aging, as GTE's management and defense attorneys argued.

Rohr continued calling Mandy Hawes for support and advice, and began to realize that there was reason to think that toxic chemicals had caused the "subjective" symptoms as well as the cancer. Hawes referred Rohr to Sheila Conrad of PHASE, the Santa Clara County Project on Health and Safety in Electronics, in Mountain View, California. Conrad sent her PHASE factsheets detailing recent research into the health effects of chemicals commonly used in electronics.

The PHASE materials, assembled by a working group of public health and labor activists, were a revelation. The PHASE materials cited research, some done up to twenty years before, linking solvents, gases, and heavy metals to neurological damage and disruption of major bodily systems: heartbeat and circulation, breathing, kidney functions, and the endocrine system of hormonal regulation. Here was the key to understanding many patterns of complaints.

As the first twenty suits gained press coverage, Fritz Hannah, human resources manager, and Jack Lacy offered to Rohr, according to her, the explanation that the symptoms were psychological in origin and not related to the work. The first forty-three plaintiffs had seen an average of ten doctors each during their years at the plant; only twelve had had the same doctor for at least five years. This background of multiple doctors made getting records expensive for Rohr (most doctors charged fees for copies, some as much as $5 per page) and made tracing the continuity of symptoms difficult.

In addition to denying the legitimacy of the workers' symptoms, GTE's defense attorneys charged that Rohr was coaching the plaintiffs in a standard list of symptoms. Attorney Carlos Martinez noticed, "The lists of symptoms on the Interrogatories were different at first, but at one point they suddenly started being the same. At one independent medical exam we set up, the doctor, by mistake, read one plaintiff the list of symptoms claimed by another plaintiff. She agreed to them." However, plaintiff files reveal that the notes in Rohr's

handwriting from the plaintiffs' initial interviews reflect varying orders and totals of symptoms. Plaintiffs report that Rohr
simply asked them what they had, and they often had similar
problems. Rohr did generate a checklist of health problems to
use when reinterviewing plaintiffs a year or more after their
first interviews, and use of the checklist may have contaminated the self-reported symptoms for some plaintiffs.

On the other hand, Rohr suspected GTE of suppressing,
"losing," or altering medical records. Carlos Martinez was,
like Rohr, in the middle. His client would deny having certain records Rohr requested, and he would have to go back to
GTE to ask again when a plaintiff produced her own copy of
the requested type of record. Martinez wrote to Rohr:

> At this time, we have no master file of forms used in an
> individual medical file nor do we have a master policy
> which indicated when certain forms were used, why cer-
> tain forms are marked occupational and others are marked
> nonoccupational and why for a certain period of time Jack
> Lacy instructed the employee medical service not to write
> down the type of complaint [when the worker] appeared in
> employee medical services. We cannot explain why Ellen
> Kayser or other personnel's medical records appear to be
> cut and pasted, why sometimes entries are written all in
> one handwriting and appear to be made at the same time.
> It may be that at one time they attempted to transfer infor-
> mation from the daily log into the individual employee's
> file and that this was discontinued for some reason. After
> a thorough investigation, we have no reason to believe
> that any medical records were intentionally deleted or
> changed. . . . As you can now tell, most employees were

in the employee medical services on several, if not dozens more occasions than their individual files previously indicated.

Catch 22: Do You Remember Losing Your Memory?

Rohr had every right to feel she had stepped through the looking glass as the number of clients grew. She knew from the PHASE material that toxic chemicals could cause memory loss, mental impairment, and emotional problems, which made it very difficult to evaluate some people's stories as they came in for interviews. If the woman had a sterling work and family history and seemed deeply ashamed of episodes of falling and forgetfulness, it was one thing. But others seemed so . . . *batty*. After repeated interviews with one plaintiff who talked with the spaced-out lisp of a lost little girl, Josephine told the woman's cousin, also a plaintiff, "I don't know if we can file her case! How would anybody believe her? She's such a dingbat!"

"Yeah, she's always been a dingbat," the woman replied. "But she's much worse after working in that job."

Whether or not they were all attributable to chemical effects, the range of incidents Rohr heard was bizarre. One woman said, "My kids laugh at me. I used to keep a perfect house, and now they'll find the sugar bowl in the freezer. I put it there." Several women related getting lost on the freeway, which is difficult to do in Albuquerque because there are only two. Three had had the same experience—having to pull over and wait to remember where they had set out to go. Some of the stories came from the husbands and children: Mom is so *out of it* lately. Marjorie Padilla said, "It's a joke any more; my friends say, 'Let's go over and watch Margie do the dry heaves.' "

Several had been experiencing auditory and visual hallucinations—hearing their names called when home alone, hearing footsteps and seeing moving shadows and things crawling on them. Two women had had Native American healers come to bless their houses because of these symptoms, and had told no one out of fear of being considered crazy. Others told of recurring nightmares; one woman dreamed several times that she heard footsteps approach her bedroom door, which was then opened by a man with no hands who reached out to her. Other workers related this dream to a horrifying accident that happened at the plant in the mid-1970s, when a worker named Vince Montaño had had both hands cut off.

Sometimes Rohr doubted one client; sometimes her daughter or son scoffed at another. Often she stayed at the office late into the night, drinking coffee, smiling and shaking her head at one story, then staring out the window and wondering.

Managing a Growing Group

As long as the total group numbered fewer than fifty, Rohr was able to maintain personal relationships with them, and they were all able to share some intimacy and social support. Rohr had a higher cash flow than they did, but was also in continual financial crisis. In 1985, she took a second mortgage on her house for $30,000 and still lost $10,000 that year. She was getting no money from the dozens of GTE plaintiffs, since attorney's fees for worker's compensation are awarded only upon settlement of the case. The clients had no money anyway. She began considering applying for food stamps.

During 1985, Rohr's son, Rob, twenty-five, had come to work as her office manager after working as a legal clerk in the army. Her daughter, Claudia, twenty-three, also joined the office to replace the receptionist whom Rohr could no longer pay. There was some advantage in employing one's children, beyond being able to pay them in room and board; the office took on an informal family air. The plaintiffs who had no legal experience were a little more comfortable when Rob or Claudia reassured them that "Mother will be right with you."

As the list of accepted plaintiffs and prospective clients grew, the office began to be overwhelmed. The appointment book was filled with new client interviews, depositions, and hearings. Rohr spent hours each day at sites chosen by the defense, taking depositions of plaintiffs; when she returned to the office, the small waiting area was filled with people waiting to see her. Some were being dunned for past-due medical bills, some had remembered something she had left out of her interviews, and some were there for the first time.

Scores of ex–GTE workers came in for interviews, including "flakes" and others repeating memorized lists of similar symptoms. Josephine accepted as clients only those whose accounts of their work experiences and health problems had an original ring or could be verified by other plaintiffs. Still, Josephine, Claudia, Rob, and their friends had to devote Saturdays and Sundays to marathon triage operations on the paperwork, especially after documents began to arrive from GTE's defense team in cartons, hundreds of pages sometimes in no apparent order. Rohr had asked GTE to produce employee personnel and medical records, chemical lists, Material Safety Data Sheets from the chemicals' makers, process descriptions, blueprints of the plant, safety regulations, and many other documents. GTE denied having several types of materials, until plaintiffs produced their own copies and forced GTE to look harder. Rohr's office often looked like a disaster area. Things were lost; deadlines were missed. She and her kids got some relief by giving black humor free rein, joking about the flaky clients, the button-down GTE attorneys, the mess in the office, the junk food they were eating.

An impending deadline in the Immigration and Naturalization Service's amnesty program caused an upsurge in Mexican clients unrelated to the GTE cases, seeking residence permits and amnesty petitions; whole families would show up at the office after they got off work—which was also after Rohr's office hours—and all day Saturday and Sunday. Rohr tried stoically to help them, often underestimating the time these cases would take. Her accounts-receivable ledger for 1985 and 1986 shows many monthly $25 and $30 cash installments for immigration services.

Josephine began to see a psychologist with an office down the hall from her own, handling his mother's estate papers in trade for therapy. Lenkurt had five attorneys assigned to the case with unlimited clerical help, two full-time people to make copies, and a runner to deliver documents around town. The GTE defense firm occupied extensive uptown offices and, just before settlement negotiations intensified, they reserved one quarter of another floor in their office building for the GTE case, with a network of computers to keep track of documents. Rohr and her family worked out of three small rooms, bringing food in from a convenience store and fast-food restaurant next door.

The average workers' compensation case in New Mexico

(filed in state district court, before a board system was created in December 1986) usually required a maximum of nine pleadings—the first filing, notice to defendant, requests for production of documents and evidence, motions for discontinuance or compelling the production of documents, and then a decision by the judge. The GTE cases reached 191 pleadings, a numbing crossfire of motions, third and fourth requests for production of documents, accusations, and explanations. The ultimate total of documents generated by both sides surpassed 100,000 pages. Technically, they were still separate workers' compensation cases, but gradually evolved into a de facto class-action tort case claiming similar exposure and causation factors for multiple plaintiffs. The case came close to swamping Rohr and her staff.

Woody Smith, the district court judge assigned the case, ruled in GTE's favor on a string of initial motions to dismiss plaintiffs' claims with prejudice for being beyond the statute of limitations for filing or because the plaintiffs were still employed. Smith ordered Rohr to respond more quickly to requests for documents. Rohr says,

> Normally you go into court with your case done, your investigation complete and your experts hired. But I had none of those done. It was unbelievable! My inexperience! My bungling! I was like a bull in a china shop. I had not tried even one case before, and Carlos was eating me alive!

After reading more of the plaintiffs' depositions, Judge Smith began to give Rohr more time. He said,

> Fairness has to come before the law sometimes. When a law stands in the way of being fair, you have to find a way

around it. I saw right off that she was in over her head. And then some of those depositions . . . like that one about the "mood swings" [*he chuckled and shook his head, smiling*]. But after I read more, you couldn't fail to believe some of these people had really been abused. It was too weird. The depositions made me decide that this case was important enough that it needed to be heard, not dismissed on technicalities.

Rohr battled her feelings of being swamped by getting advice from Mandy Hawes, the Silicon Valley attorney: "She used to call me every week to encourage me. She really kept me from going crazy for a while."

In June, Hawes arranged for a visit to New Mexico by four delegates from Disabled Workers United. Two of the delegates were victims of toxic overexposures and had experienced symptoms similar to the Lenkurt workers. The four stayed in Albuquerque the first week of June. There were sixteen ex-GTE workers at their first meeting, and thirty of the forty-four current plaintiffs at the second. Anita Zimmerman, one of the visitors, had been a nurse before going to work in production for Advanced Micro Devices (AMD) in Sunnyvale, California. AMD was one of the most celebrated Silicon Valley firms, offering in 1980 a companywide lottery awarding $1,000 a month for twenty years, called "The American Dream." Zimmerman lost her voice during a chlorine gas leak at AMD, and then got acute bronchial asthma after the leaks continued for six weeks. At the meetings with Albuquerque GTE workers, Zimmerman described an outpouring of frustration and emotion; she says there was bonding between the workers and an eagerness to learn how they could turn personal experiences into collective action. Rohr said the visitors had done five months' organizing for her in eight days.

After these meetings, Yolanda Lozano, the third plaintiff to join the case, emerged as the worker spokesperson. She held "victim support" meetings at her house. The local PBS television affiliate's public affairs commentator, Hal Rhodes, interviewed Yolanda at home, building two investigative programs around her testimony.

Death and the Cold Reality of Corporate Litigation

In the late fall of 1985, a year into the case, the pace of events suddenly increased. On November 7, Amy Cordova Romero, age thirty-eight, died in a bedroom of her brother's house. The funeral was held in Santa Rosa, 120 miles east of Albuquerque; her family asked Josephine to come. GTE's defense had filed a request for summary judgment (asking Judge Smith to dismiss all cases as unworthy), and the hearing had been set for early morning on the day following Amy's afternoon funeral. Josephine wanted to go to the funeral and stay overnight, for fellowship with Amy's family, whom she had never met, and because the return drive could be treacherous at night, crossing the Sandia Mountains through Tijeras Canyon, famous in the winter for its crosswinds, icing, and jackknifed semitrailers. She asked Carlos Martinez to request a one-day postponement; he responded, "My clients won't let me."

After missing Amy's funeral to fight GTE's efforts to have all suits dismissed, Rohr hit a wall of doubt and exhaustion as Christmas approached. She had remortgaged her home, spent the money, and was sinking deeper into debt. Her campaign against the world's fourth-largest utility company looked as bleak as some had warned her it would.

I began thinking I was getting nowhere, wasting everybody's time, and meanwhile people were getting sicker and
dying. I told Claudia I was going to quit. Everybody had
been right; you can't win against a GTE. I stayed at home
in a daze, feeling like a fool and a failure and dreading
Christmas because we didn't have any money, and trying
to figure out how I could tell everybody I was quitting.

About a week before Christmas, two plaintiffs, sisters in their
fifties who had never married and lived together, knocked
on her door. It was Millie and Jackie Glawe, on a rare trip
to town from their house in the Jemez Mountains northwest
of Albuquerque. Both had quit after being on medical leave
from GTE. Millie had had ovarian cancer. She and Jackie had
worked beside each other for thirteen years at the plant, with
near-perfect attendance. They were very quiet, shy people.

They were standing there on the porch. They had come to
give me some gifts. They had hardly any money at all, but
they had wrapped this little clock and this little religious
figure. I was so touched. I mean, I broke down, we all three
did. I knew then that I couldn't quit. How could I desert
these people?

After Christmas, a law school classmate of Rohr's who was
subletting a cubicle in her office relayed to Josephine an offer
of help. The woman's husband's brother-in-law was an attorney who would perhaps be interested in joining the case. He
offered courtroom experience and could afford to put upward
of $300,000 into the case. And so at the end of the case's first
full year, Josephine had found a colleague to help with the
daunting paperwork and deadlines, and a promise of money.

In the absence of government regulatory oversight and worker input into plant safety operations, freelance lawyers and networks of activists were the only source of sustained response to the GTE worker health complaints. Mandy Hawes says,

Workers are a phenomenal source of information on workplace hazards. The most workers from any one plant I've had is seven; with so many from one place, Josephine stands a good chance of having a fact-finder see the picture more clearly. There are virtually no studies done; as odd as it may sound, Josephine and I, as lawyers, are actually gathering medical and toxics data by default.

Medical Experts:
Interpreting Facts

Josephine Rohr could joust with GTE Lenkurt's legal team only so long before needing infusions of money and trial preparation experience. The greatest single expense in legal controversies over medical issues is the hiring of expert medical witnesses to interpret the data, carry the argument, and impress judge and jury. Both GTE Lenkurt and Josephine Rohr went outside Albuquerque to find expert witnesses on the national market. GTE retained a Brookline, Massachusetts, firm with experience in large toxics cases, Epidemiology Resources, Inc., which coordinated all expert scientific testimony for the defense. For Rohr, on the other hand, the process entailed chance connections with a series of unaffiliated professionals from across the country who managed to mount a helter-skelter case that continued to unfold right up to the scheduled trial date.

Recruiting Experts from Silicon Valley

In 1986 Mandy Hawes arranged for Rohr to attend an invitation-only three-day conference on workplace toxics held at a monastery near San Jose, California. The workshop's sponsors, a labor-left coalition, felt a need for tight security, since the site was so close to the hundreds of corporations in Silicon Valley; so they billed the conference publicly as a "marriage encounter" to avoid attracting spies from corporations anxious to defeat health suits against themselves. Those invited included Rohr and Hawes, the only attorneys; Nancy Lessin of Mass-COSH (a labor-oriented Committee on Safety and Health from the Boston area); Randy Wilson of the AFL-CIO; labor organizers from North and South Carolina, states experiencing a "Silicon Valley fever" interest in high-tech business and research; professors of epidemiology and toxicology; and several electronics industry workers.

At the conference Rohr met Bob Harrison, director of the Occupational Medicine Clinic, University of California at San Francisco, and James Cone, chief of the Occupational Health Clinic at San Francisco General Hospital. Harrison and Cone had been interested in the effects of toxic chemicals on workers for some time, and had just collaborated on a research paper describing workers with multiple sensitivities to chemicals. Rohr described the GTE plaintiff group and asked the two physicians if they would be expert witnesses when she got the case to trial. The two informally agreed.

In spring 1986, the attorney who had offered to help Rohr accompanied her and six of the most seriously affected plaintiffs to get medical evaluations from Bob Harrison and James Cone. These six workers had been examined in Albuquerque by William Wiese, director of community medicine at the University of New Mexico School of Medicine, who found no pattern of illness or symptoms he could identify. Harrison, Cone, and Rosemarie Bowler, a Bay Area psychologist, also did not find dramatic enough clinical or laboratory evidence to justify strong testimony in court. Harrison says,

On the basis of the six we examined, we felt there were other psychological factors—anxiety, depression, family and personal disruption. They needed counseling and group therapy. Our tests are not sensitive enough to discriminate organic solvent effects from ambient functional problems. Depression can mask the chemically caused symptoms. It's a very common problem and very hard to differentiate the two. Both are work-related, but you need abnormal EEGs, blood tests, or something to convince judges and juries.

Rohr was crestfallen that Harrison and Cone could not come to firm conclusions that would stand up in court. By the same token, Harrison, Cone, and Bowler became disillusioned with Rohr when it took them nearly a year to get paid their examination fees. Rohr's associate had taken on the responsibility of corresponding with, and paying, the experts, as well as preparing to take to trial the cases that were most ready. Rohr continued to spend her time interviewing new plaintiffs, filing and answering pleadings, and being the emotional hub of the plaintiff group. She had little contact with her associate, and he had little or no contact with plaintiffs.

After the disappointment with Cone and Harrison, the associate began casting around for other contacts. He brought in attorney James Riley, with whom he had worked in Albuquerque. Riley was now based in Boston, and after flying to California to interview Harrison and Cone again, he decided to try to get Alan Levin, an immunologist and veteran plaintiff's expert witness. Riley knew him through Levin's testimony in the well-publicized cases of childhood leukemia clusters, blamed on well water contaminated by solvents, in Woburn, Massachusetts, in the early 1980s. Rohr and Riley went to San Francisco in March to meet with Levin. Levin required a retainer fee of $10,000; Riley could afford that and wrote a check. The GTE plaintiffs now had a new and controversial expert witness.

The Inescapable Politics of Medical Research and Expert Witnesses

Associating with Levin assured the plaintiff team of a courtroom debate not only about their clients' conditions, but also about Levin, whose theories about chemically caused immune system dysregulation have been one of the flashpoints of toxics litigation since the early 1980s. As a vocal plaintiffs' witness in several cases of chemical exposure awarding high tort damages, Levin has attracted organized opposition from the chemical industry. His impact on judges and juries is so strong that the industry has organized seminars in how to refute his testimony.

Board certified in allergy, immunology, pathology, and clinical pathology, Levin is a fellow of the American College of Emergency Physicians, member of the Society for Clinical Ecology, the American Medical Association, and the California Board of Medical Quality Assurance. His personality combines sarcastic iconoclasm, avant-garde medical theories, and a combative self-confidence. According to Levin, his gross income in 1986 from his thriving practice in immunology and allergy in San Francisco was $900,000, of which 10 to 15 percent came from expert-witness fees in court cases.

Levin says that organized opposition to his testimony against the chemical industry has become Kafkaesque. He claims that he was subject to, in the same week in 1985, his fourth straight IRS audit; an investigation by the Food and Drug Administration; charges by the California Board of Medical Quality Assurance (on which he sits) that he was having sex with his patients in his office; and charges by the Veteran's Administration, resulting from an audit of his files, that he had fraudulently accepted $40.78 in overpayment in 1970. Levin finds it no coincidence that these probes came as he was testifying against the Velsicol Chemical Corporation in a multimillion-dollar suit in Tennessee.[1]

According to Levin, the ties between the chemical, medical, and aerospace industries are historically strong and rapidly becoming monopolistic. It is his theory that aerospace and defense contractors, faced with an inevitable end to the cold war and thus declining profits from weapons manufacture, have been positioning themselves for several years to take over medical research and development. He points out that McDonnell Douglas makes 60 percent of the hardware used in hospitals; that Lockheed owns the Dialog database, the dominant software of medical research; and that Hughes Aircraft, with huge investment in medical research endowments at the University of California at San Francisco, Johns Hopkins, and Harvard, will have a greater budget for medical research than the federal government by the year 2000. Levin also finds conflicts of interest in the support of mainstream medical research by drug manufacturers—for example, in the fact that 83 percent of the operating budget for the *New England Journal of Medicine* comes from advertising placed by drug makers. Levin is certainly not the first to point out these links. In his case, his experience as a Navy doctor in the Vietnam War gives his criticisms of military-medical ties a personal urgency.

During the Tet Offensive launched by North Vietnam against the South Vietnamese and their American allies in 1967, Levin was a naval flight surgeon attached to the Marines.

The 1st Marine Battalion, 9th Regiment, 3rd Division was ordered in May 1967 to take Hills 881 and 861 near Khe Sahn, just after being forced to exchange their M-14 rifles for new M-16s, manufactured by Mattel, the toy manufacturer, which had failed many field tests. When the Marines encountered a larger force of North Vietnamese, many of their M-16s jammed and they were slaughtered. Levin recalls,

> There were 700 bodies in eighteen hours, 1,200 in two days. The unit's original doc went crazy, so I processed the dead myself. I zipped up 700 body bags in one day. A general, William Green, came when it was over and wanted to decorate me. I popped off, "Some of us would rather see our guys have a fighting chance to live than get complimented for burying them." He threatened to court-martial me.

Levin proudly claims two other threatened courts-martial for insubordination. He says he wanted to be an astronaut, but ended up doctoring the dead on battlefields.

Levin is driven to take up arms against what he considers idiots, profiteers, and heartless industrialists everywhere, although it can hardly be said that there is no profit in his own participation in toxics litigation. Like Josephine Rohr, he came to the GTE case with the personality of an outsider and with strong antiauthoritarian politics, a stance that he maintains not only against the military but against modern medical research and publishing.

Levin alienates many in the medical world with his sarcasm. In 1986–87, he wrote a parody of medical database support of causation theories in which he "proved" that baseball bats cannot cause skull fractures. It is obviously an allegory for academic criticism of his own chemical-exposure theories. In a guest editorial in the newsletter of the American Association of Environmental Medicine, he created a snide medical school theoretician who ridicules a general practitioner for believing a patient who tells him his headache was caused by being hit on the head with a baseball bat the previous day. The clinician has found a skull fracture by X-ray, but the academician proves the falsity of the association by searching the Medline database of the Index Medicus literature using the descriptors "skull fracture," "headache," and "baseball bat" and finding no correlation in thousands of citations.

Levin sees a dichotomy between practicing doctors, like himself, who encounter patients with symptoms related to life events, and isolated, laboratory-oriented researchers. In an interview, he stated:

> Academicians set standards of care without knowing anything about the course of disease expression by a variety of people. People who write papers don't see patients, and vice-versa. I'm aligned with the overwhelming majority of practicing physicians in the opinion that knowing disease means knowing *people,* not lab studies. . . . It was practitioners, not researchers, who were the first to discover the links between ionizing radiation and cancer, asbestos and mesothelioma, and DES [the drug diethylstilbestrol] and cancer of the female reproductive tract. In fact, we owe the discovery of the link between DES and adenocarcinoma of the uterus to the chance meeting of two clinicians who happened to be trapped in the same broken elevator.

The consequence for the study of toxic substances, according to this argument, is that laboratory calibration of the parts per billion of a chemical that are dangerous to a few healthy research subjects under controlled conditions is meaningless and should not be used as a basis for legally permissible levels of exposure to workers under other stresses at the same time. The standards will have to be changed when other subjects express different reactions to the substances. Toxicologists join Levin on this point.[2]

Levin's interest in the effects of chemical stressors on the immune system began in 1979–80, when an allergist friend from Tennessee testified in pretrial motions of the Velsicol trial. Velsicol was the manufacturer of Chlordane and Heptachlor, pesticides used in the war on termites. Between 1964 and 1973, when the state of Tennessee ordered the corporation to stop, Velsicol had buried 200,000 drums of pesticide manufacturing wastes in a remote area inhabited by several dozen families. One well, serving six families, was found to contain 2,400 times the permissible level of carbon tetrachloride, known to cause cancer and diseases of the kidney and liver. In 1986, a Tennessee court awarded a group of residents $5 million in medical and property compensation and $7.5 million in punitive damages (the award is on appeal and subject to reduction by appellate judges). The trial judge castigated Velsicol for its "gross negligence and wanton disregard for the health and well being" of area residents.[3]

Trials like the Velsicol case have provided money and a dramatic platform for the debate of theories about chemicals and the immune system. Human immune response was elevated to the top of research agendas by the AIDS controversies of the 1980s. Thus toxic tort litigation and immune system theories provided Levin with a prominent stage.

In 1986, another trial in which Levin had testified for the plaintiffs resulted in a large out-of-court settlement. Eight families in the Woburn case settled for $8 million, and in another case involving chemical contamination, a Missouri group got a $19 million settlement.[4] There is a huge case pending in Tucson, where over three hundred plaintiffs claim various illnesses because of trichloroethylene leaked into well water by Hughes Aircraft. Levin expects to take part on the plaintiffs' side in the Hughes case.

The theory being debated in these trials is based on accepted knowledge of the human immune response, but goes beyond accepted diagnostic conclusions. In the body's response to perceived invaders, there are several cellular agents. "Helper" (T4) cells activate other white blood cells that attack invaders. After the invading cells have been controlled, "suppressor" (T8) cells turn off the fighting response. Normally, there are twice as many helper cells as suppressor cells, but weakened immune systems can have twice as many suppressors as helpers. According to Levin's theory, this can weaken the immune response so much that any organ system in the body can be attacked successfully. With the help of test-tube immuno-assays developed by his wife, the internist Vera Byars, Levin has come to champion the theory that measuring the ratio of T4 cells to T8 cells in the blood gives an accurate measure of how much damage has been done to a person's immune system.

Levin's theory holds that petrochemicals, formaldehyde, sulfur compounds, household cosmetics and cleaning compounds, vinyl plastics, polyester, petro-combustion products, and other chemicals can enter the body, combine with natural proteins, overstimulate or weaken certain critical cells in the immune system, and create "chemically induced immune system dysregulation." The victim may manifest a wide spectrum of symptoms—coughing, increased colds and flus, extreme allergic reactions to chemical compounds, headaches, cancer, even mental illness. Levin defines the immune-system dysregulation theory as "about at the point where radiation research was in the 1940s." He and Byars describe the syndrome as "environmental illness," which demonstrates biochemical mechanisms "quite similar to the effects of ionizing radiation. . . . In diagnosis, one is struck with the similarity between the symptoms of environmental illness and the symptoms of acute infectious hepatitis." [5]

Some working in occupational health and studying electronics workers disagree. Robert Harrison, who had conducted examinations on some of the plaintiffs, finds Levin's theory "scientifically unproven. At the same time, I believe the work of investigating these workers' complaints is good and important."

For many physicians who distrust the legal profession, Levin has tainted his theory by pursuing it in the courts before submitting it to peer review in prestigious medical journals. Levin argues that lives are being ruined and lost while conservative debate plods along in the literature. There is public pressure to win lawsuits, and no shortage of plaintiffs.

During 1986–87, Levin flew his plane to Albuquerque several times to do marathon examination sessions on the GTE plaintiffs. He saw each plaintiff for perhaps twenty minutes, taking personal and family histories, drawing blood, and doing a physical exam. He supplemented these examinations by reviewing plaintiffs' medical records in Rohr's office. He wasted no time coming to a conclusion. When I asked him in March 1986, just after his first visit to Albuquerque to examine plaintiffs, what he thought had happened at the Lenkurt facility, he answered, "Auschwitz!" His apocalyptic style induced Rohr to tell plaintiffs, "Alan says 80 percent of you are going to get cancer." Naturally, conservative physicians find Levin's conclusions alarming.

GTE's defense attorney, Carlos Martinez of Albuquerque, conducted twenty-four hours of deposition with Levin in preparation for trial. By order of the court, the defense paid Levin for deposition testimony—$8,000 for two days in Albuquerque. Martinez took issue with Levin's glib vagueness. For instance, Levin had said that GTE chemicals gave Emma Leyba breast cancer, and when Martinez pointed out that her mother and sister had both had breast cancer, indicating a strong family predictor of the disease, Levin just said that the chemicals made her get it quicker. On another occasion Levin cited T-cell ratios as evidence that another plaintiff had chemical damage, and when Martinez pointed out that some of the nonexposed control group's ratios were wider apart than that plaintiff's in question, Levin shrugged and said that it didn't matter because he relied on her physical exam also. Levin's argument is difficult for defendants because developments in tort law have shifted responsibility for demonstrating proof from plaintiffs to defendant, and proving one's corporate innocence is difficult at this stage of knowledge about chemical toxicity.

GTE's Medical Experts

Alexander Walker, who coordinated GTE's defense for Epidemiology Resources, Inc. (ERI), spoke as the principal expert for the defense and described, in his deposition, the roles of other ERI staffers and subcontractors. While his conclusions essentially absolved GTE of any responsibility for plaintiff illness, his defense was mild, conditional, and left several issues open to dispute. His testimony faced the fact that it is extremely difficult to understand what is happening in complex cases of toxic exposure, using diagnostic and statistical methods that are not only expensive but rapidly changing. He acknowledged that epidemiology is a statistical exercise demanding, for its more reliable conclusions, more money, time, and data than he had for the GTE case.

Epidemiology Resources, Inc. had a prominent role in the Manville Corporation's defense against lawsuits from asbestos workers dying of asbestosis and lung cancer in the late 1970s and early 1980s. According to Paul Brodeur, who has studied the asbestos controversies for twenty years, Alexander Walker and ERI provided key testimony that helped Manville file for reorganization under Chapter 11 of the federal bankruptcy statutes, leaving the corporation's assets protected.[6]

Brodeur reports that Walker acquiesced to pressure from a lawyer working as "litigation risk analyst" for Manville in 1981. Manville wanted to know how many suits it might face in the future. Irving J. Selikoff of Mt. Sinai Medical Center's Environmental Science Lab, the world's leading expert on asbestos disease and an outspoken worker advocate, had estimated 270,000 excess deaths from asbestos-related disease between 1980 and 2010. Depositions show that Manville's analyst asked Walker to choose the lowest possible figure he could; Walker, discounting Selikoff's risk multipliers for smoking combined with asbestos exposure, came up with 139,000 excess deaths, one half of Selikoff's estimate. Dr. Nancy Dreyer, founder and president of ERI, determined 180,000 excess deaths. A Dr. Nicholson of Mt. Sinai criticized Walker for omitting from his equation the hundreds of thousands of construction workers exposed to open spraying of asbestos between 1958 and 1972. Asbestos was used at that time to coat pipes and boilers for insulation.

Brodeur concludes that this insistence on the lowest possible scientific estimate—a political choice—enabled Manville to predict enough future deaths (some of which would undoubtedly result in lawsuits) to justify filing for reorganization under Chapter 11, but not so many as to force liquidation under Chapter 7. Manville, with $2 billion in assets, was the healthiest corporation ever to file under Chapter 11, and Brodeur implies that they cheated many potential litigants out of compensation due. He implies that Walker and ERI were used by corporate attorneys to protect a corporation against whom juries had awarded, by 1981, $1,500,000 in punitive damages, over and above compensation awards, for "outrageous and reckless misconduct." These punitive amounts had been awarded after fewer than 40 of the 7,000 lawsuits against Manville had gone to trial.[7] Brodeur provides no conclusive argument that liquidating Manville's assets under Chapter 7 would have been the better way to guarantee future plaintiffs compensation; he assumes that an ethical response to Manville's "outrageous misconduct" requires a punitive liquidation.

Alexander Walker is a Harvard-trained physician who did residency in pediatrics and neurology. He is not board certified in any field and concentrates full time on the mathematics of study design for epidemiology, which is the study of disease occurrence in populations.[8] Retained by GTE in the spring of 1985, he seems to have done a careful job of reviewing the limited materials supplied him by GTE's defense firm. He examined no plaintiffs. His deposition is highlighted by explanation of a serious flaw in the lab results on Levin's plaintiffs' blood samples, and by Walker's admission of the limitations inherent in doing epidemiology on groups like the GTE workers.

Walker found that two laboratory analyses of the blood

samples Levin took from the plaintiffs found differing counts of white blood cells and lymphocytes, differences of a magnitude not compatible with chance variance. Two samples were taken from each plaintiff; one sample was analyzed by a lab in Albuquerque and the other by a lab in Oakland. Walker says,

> I assume degradation of sample en route, an actual loss of white blood cells between the two points. I think, biologically, it casts doubt on further refined tests done at Oakland. I am the only person who considered quality control to be an issue in this case. So it gives me a lack of confidence in the analytic capacity of persons who relied on one or the other samples without even bothering to compare whether or not they were consistent with one another.

These were the blood samples Alan Levin used to determine the ratios of helper to suppressor cells, his key indicator of immune system damage. As for that theory, Walker said,

> I would certainly accept the proposition that certain of the chemicals used at Lenkurt in the period we're considering can cause neurological damage if the exposure is in excess. I accept the proposition that factors external to the individual—or should we say factors initially external—can affect his immune system and even damage his immune system. My disagreement with Dr. Levin is not only derived from his data and mode of reasoning, but also from the fact that the data so far as I understand them at this point are not what you would expect from a group of people with a chronic immune deficiency.

The Limitations of GTE's Expert Testimony

To come to that conclusion, Walker had to analyze the wide range of multiple, less well defined complaints reported by the workers. There were political and financial limitations on his choices of methodology:

> We didn't have any opportunity that I could see here for a cohort study that would really address the issues of disease occurrence in this population. . . . The situation we are walking into now is one in which [management has] a pecuniary interest in [the plaintiffs'] disease . . . and [is] rarely amenable to evaluation of individuals, and certainly not amenable to population-based studies, using novel techniques of ascertainment.

Walker's conclusions about the etiology of the plaintiffs' symptoms are based on his review of the records of only twenty-four of the plaintiffs, a review of the chemicals used at the plant, and a review of dispensary logs from the 1970s. From 1971 to 1979, GTE Lenkurt nurses had workers enter their own complaints in spiral-bound notebooks when they visited the dispensary. These haphazard records were never copied and entered into the workers' files, which may explain why many plaintiffs who worked throughout the 1970s and visited the nurse often have little or no record of such visits in their files before 1980. One existing notebook for December–January 1975–76 showed 945 dispensary visits. Walker had an ERI employee analyze and summarize these incomplete log entries, and concluded that, since he could see no patterns of mass episodes of headache, nausea, or dizziness, there were no short-term exposures to high levels of toxics. He ruled out toxic-related respiratory effects because a seasonal pattern of colds and flus *did* emerge, suggesting to him normal patterns implicating weather and community-wide viral infection.

Walker identified the common theme among the twenty-four plaintiffs he reviewed as "corresponding to the symptoms which have no explicit documentation in the medical records . . . those being headaches, sometimes nausea, sometimes the paresthesia . . . those, as we've discussed, are symptoms which have in the past been related to excessive exposure to . . . a variety of solvents." In his deposition, he said that chemical causation of these and associated symptoms and complaints could not be ruled out for eighteen of the twenty-four workers. For several workers, he dismissed chemical causation for colds, asthma, hypertension, heart palpitations, and psychological problems, all of which are reported as effects in the solvent literature since 1984.

Walker attributed many symptoms to plaintiffs' obesity or smoking and avoided acknowledging partial chemical causation. That is, he implicitly defined obesity and smoking more sharply as *sufficient* causes and chemicals as *necessary* causes. This bias is inconsistent with epidemiologic understanding of multiple causation, as described by Kenneth J. Rothman's standard text:

> [A cause is] any event, condition, or characteristic that plays an essential role in producing an occurrence of [a] disease. . . . The extent of biologic interaction between two factors is, in principle, dependent on the relative prevalence of other factors. All disease is caused by interaction [and] . . . it is a fallacy to try to assign proportions to the various component causes. In a causal constellation, all the components are necessary causes of *all* the disease.[9]

This spotlights the compromises made when medicine and law are yoked: "pure" epidemiology takes a holistic view; epidemiology in service of one side in an adversarial legal fight makes political decisions to inflate or ignore one factor or another. Like Solomon with his sword, law seeks to divide the limited resources cleanly, if arbitrarily, making commodities of the parts of the body, the roles in a life, and the relationships between people.

Walker left open the debate about poor ventilation and inaccurate chemical exposure standards. While saying, "There is a strong indication that the ventilation system performed properly," he admitted that "it is not possible to evaluate how effective the system was in actual use because data on the parameters such as rate of contaminant generation, process temperatures, work practices and equipment location relative to ventilation inlet location were not known." Regarding threshold limit values (TLVs), set by the American Council of Governmental and Industrial Hygienists, and permissible exposure levels (PELs), set by NIOSH, Walker admitted:

> I learned in connection with this particular case that for many of these subsets the best available data is nonetheless rather limited. I had thought prior to this case that setting these limits had a more thorough-going scientific basis than it actually does. . . . We don't have any clear information for most of these [chemicals] at what levels, if any, there would be permanent, as opposed to transient, health effects.

Thus the expensive medical experts presented the judge with complex testimony, the believability of which seemed to depend as much on the ideology and personality of each man as on independently verifiable references. They could hardly have represented insider–outsider positions more dramatically. However, Walker's studied caution seemed to highlight the limitations of his surveys of the GTE workers, while Levin's extravagant conviction seemed to harmonize with the workers' own shocking stories in their depositions.

Additional Help for Rohr, and a New Legal Phase

In September 1986, GTE Lenkurt sold the Albuquerque plant to Siemens, an electronics conglomerate based in Germany. Lenkurt and Siemens officials stated to the press that the sale of the plant had nothing to do with the health litigation, which had just passed its second anniversary at that time.

In December, *CBS Evening News* ran a story on the case against GTE. A viewer in West Texas took special notice: he was Roger Jatco, an attorney who was pressing a case for several aircraft mechanics who claimed to have been disabled by working with petrochemical solvents. Jatco needed an expert witness; Rohr gave him Levin's telephone number. By this time Rohr knew that her evidence justified a second type of lawsuit: charging the chemicals manufacturers with selling hazardous products to end-users like GTE. Plant records identified the chemical suppliers as Dow, DuPont, and Shell, among others. As overwhelming as the workers' compensation cases had been, Rohr saw that "product liability" suits against the chemical multinationals would be far bigger and far more expensive.

Around the first of March, 1987, Jatco called Rohr from

Texas to chat and found her despondent. Her local colleague did not want to continue into the products liability suits and did not seem to have the money he claimed he would. Trial was approaching, and without solid financial resources, it would be futile to try to go into court. Rohr told Jatco, "I'm going to have to give away my cases. I can't do it by myself. I'll have to offer them to one of these big national firms that specialize in toxic injury cases. I hate doing it, because the women will just become numbers then."

Jatco came to Albuquerque to assess the case and, considering it "fantastic," offered to find Rohr major support in Texas. He had known plaintiffs' lawyer Pat Maloney of San Antonio for twenty years, and Maloney had just won the largest award ever given a plaintiff in the United States—$27.5 million, in a case involving a man whose family was immolated in a propane truck explosion. Jatco soon called Rohr to confirm that Maloney would fund and direct the product liability case, as well as keep the occupational disease cases afloat until they reached trial or settlement. He assured her that Maloney would not let the plaintiffs become mere numbers.

Jatco came to Albuquerque to help Rohr prepare for trial. Looking like a tall version of Steve McQueen and talking in a smooth, burry smoker's voice, Jatco brought a sense of cool authority to the increasingly hectic proceedings. He charmed the plaintiffs, served as intermediary between Maloney and Rohr, and did not intrude on her position as emotional leader of the plaintiffs' campaign.

Settlement:
Deals for Verdicts

7 Rohr had initiated the out-of-court settlement process on September 5, 1986, the day after GTE's defense attorneys took Alan Levin's deposition. She did it by delivering the following settlement demands to GTE: she would be willing to drop the court cases if GTE would provide disabled plaintiffs with medical coverage for life; pay them the maximum benefits according to the Occupational Disease and Disablement Act, or $60,000 to $100,000 per worker; and provide rehabilitation and education for those who could still work in other jobs (which she estimated at 30 percent of the plaintiffs). Three more plaintiffs had died during 1986: Laird Wood, an instrument technician, died of cancer at sixty-three; Jenny Montaño had a heart attack at thirty-three, after which her doctors compared her athersclerotic condition to that of a sixty-five-year-old; and Carlotta Leon died of ovarian cancer at forty-four. Rohr felt she must begin pressing the issue because GTE had the resources to continue delaying while her people went without medical and financial help.

Neither plaintiffs, defendant, nor judge wanted the cases to go to court. The testimony of the experts had not proved conclusive for either side, and the judge estimated the trials of 120 occupational disease and disablement cases would take six months to a year of full-time courtroom effort, which would not only be exhaustive and expensive for both sides, but also would clog the dockets of the other district court judges who took the cases he had to forgo. And so both sides, late in 1986, began preparing simultaneously for settlement and for trial. Judge Smith facilitated the dual-track process, first by insisting on a trial date of June 15, 1987, a strategy designed to pressure the parties to concentrate on the data they had rather than search for new experts; and then by granting postponements on the eve of the trial when it was evident the parties were close to signing an agreement. A settlement of this sort would combine the cases in one mass proposal similar to class-action toxic tort or product liability tort suits, with each worker's share based, theoretically, on the severity of her disablement and the level of her income when she last worked. The outcome would be a negotiated deal rather than a verdict. The judge would become referee rather than "fact-finder."

Why Both Sides Would Settle Out of Court

Approaching trial, the plaintiffs seemed to hold some trumps in evidence and procedure, while the defendant's strength was in the limitations of the occupational disease law. Judge Smith granted Rohr's and Jatco's motion that the plaintiffs determine the order of testimony, and Rohr had decided to present medical experts and other technical witnesses first, then management testimony, then worker testimony last. This order would make the plaintiffs' personal stories the last thing the judge would hear before he made his rulings.

However, a number of plaintiff cases did not meet the requirements of "total disablement," the only condition triggering benefits under New Mexico's Occupational Disease and Disablement Act: "Disablement means total physical incapacity by reason of an occupational disease as defined in this Act to perform any work for remuneration or profit in the pursuit in which [the employee] was engaged."[1] Courts have interpreted the latter phrase to mean a worker could be eligible for compensation even if he found work elsewhere.[2] The statute defines occupational disease as one caused *in all probability* by a direct causal connection from the ordinary processes of work in the occupation. Judge Smith said in an interview with me that this case made him realize that a person might meet the spirit of the occupational disease definition although forced to continue to work out of economic necessity. One would be forced to analyze, he said, how much a person appeared to be suffering or deteriorating while continuing to work. According to the statute, Rohr felt she had only 50 to 60 strong cases among her 120 clients at the time, and many of the most emotionally damaged did not necessarily qualify. For example, a woman who loses her uterus because of excessive bleeding caused by stress and overexposure to solvents is not disabled except for childbearing. She can still work, so she has no case under the statute. Rohr saw settling as a way of gaining some compensation for those who would probably be disqualified by the narrow applicability of the occupational law.

GTE was open to settlement primarily to end the matter quickly, quietly, and cheaply. Martinez, having taken or read the depositions of all the plaintiffs, was aware of what the press could do with some of their stories. The two years of pretrial publicity had been damaging enough. GTE regional executives in Phoenix had limited Martinez and all other Albuquerque GTE spokesmen to brief, generalized denials of responsibility, whereas plaintiffs had told detailed stories to the press. In a trial of six to twelve months, the company would be constantly on the defensive.

The Source of Compensation Formulas

All settlement proposals and counterproposals were based on the compensation formula found in the New Mexico occupational disease and workers' compensation statute. The law in force at the time, like most state workers' compensation laws, provided for a maximum award to a totally disabled worker of two thirds of his or her weekly wages, as earned on the last day worked, for a maximum period of 600 weeks. The full benefits have never been awarded in New Mexico; either the percentage of wages or period of payments, or both, have always been reduced. Thus, becoming completely disabled on the job does not qualify one for a social safety net for the rest of one's life, but merely, perhaps, half of one's accustomed level of income for a maximum of eleven years. As in the 1800s and early 1900s, when workers had to sue corporations under tort law to gain compensation, workers' compensation settlements have always been controlled by insurance company claims adjusters and corporate attorneys, a consortium with its own self-interests and the power to safeguard them. Whittled-down amounts of compensation given in lump-sum form may strike injured wage workers as more generous than they really are.[3] The system hardly recognizes occupational diseases in any case; less than one percent of all workers' compensation goes to claims of workplace-related disease.[4]

The plaintiffs' first proposal embraced a list of 104 workers, 28 of whose cases had been dismissed because they were still working or could work. The 104 plaintiffs had last worked for GTE in early 1982, on average, and their average wage at termination was $5.64 per hour (GTE figures supplied in discovery.) According to the statutory formula, these workers had an average maximum potential compensation of two thirds of $5.64, multiplied by 40 hours and 600 weeks. Thus their maximum benefits (the employer's "exposure" to liability) would be $151.20 per week, totaling $90,820 each if awarded 600 weeks. Most worked 56 hours per week with mandatory overtime, but the formula is tied to a 40-hour week. In the first settlement proposal, Rohr asked for 100 percent of the formula's benefits, adjusted to each worker's actual wages at last day worked, plus medical coverage for life. Anticipating GTE's rejection of the last provision, considering how lifetime medical costs would spiral, Rohr was prepared to ask GTE to set up a clinic to treat and investigate the plaintiffs' illnesses as a compromise.

On May 7, twenty working days from the trial date, the defense delivered to Rohr's office a list of seventy-nine experts they would call as witnesses, most of them not yet deposed. It would have taken several lawyers working around the clock just to depose the new experts, not to mention the time to digest and interpret their testimony. The defense also asked to depose plaintiffs' expert Alan Levin again, though they had already deposed him for twenty-four hours. Judge Smith ruled that these new requests violated the rules of discovery and denied them.

GTE's Offer

The next day GTE answered Rohr's settlement proposal with a counteroffer consisting of seventy compensation offers of substance and forty-four "nuisance" offers of $500 each. "Nuisance" offers are designed to buy off plaintiffs who, the defense believes, do not qualify for compensation, in exchange for their signed agreement not to sue in the future under any circumstances. Under New Mexico law, plaintiffs' attorney's fees are added as a percentage of the compensation awards, so the GTE offer included attorney's fees. The total offer was $1.27 million, with $1.1 million in compensation and $161,000 in attorney's fees. The offer included no past, current, or future medical expenses, nor a plan for a clinic. The compensation amounts were distributed as follows:

Total compensation	Number of plaintiffs
$11,000 to $16,000	46
$17,000 to $23,000	2
$24,000 to $28,000	16
$29,000 to $34,000	3
$40,000 and above	1

Sixty-four of the seventy substantive offers, then, were below $30,000, or 30 percent of the statutory compensation formula. Only 5 percent were greater than 30 percent of the formula.

The GTE offer specified that anything less than full acceptance of the complete terms of the offer would constitute rejection of the offer and it would be withdrawn. Terms included:

Dismissal of all present claims and waiver of all future claims, by plaintiffs, spouses, dependents, survivors, executors, and all legal representatives, against Lenkurt, GTE, Siemens (their successors), Kemper Insurance Group, American Motorists Insurance Company, and their subsidiaries and parents.

Agreement to a press release prepared by GTE explaining the settlement and denying any guilt or liability on the part of GTE.

Agreement by plaintiffs, their families, and plaintiffs' attorney and her employees to maintain silence regarding all terms of the settlement, refuse further comments to the press, and make no derogatory public comments about any of the defendants.

Agreement by plaintiffs' counsel to refrain from filing any future suits, on behalf of these plaintiffs or anyone else, over chemical exposure or any other claims under workers' compensation, the Occupational Disease Disablement Act, or tort suits, against any of the defendants.

Agreement by signed affidavit by plaintiff medical experts not to testify against the defendants in the future, or assist anyone else in so testifying.

Destruction of all documentation plaintiffs possess about the case generated by themselves, their experts, and clients, or by GTE.

Agreement by plaintiffs not to file any intentional tort suits against the defendants and to reimburse GTE if chemical manufacturers, successfully sued by the plaintiffs under product liability tort suits, then sue GTE for negligently using their products and making them liable to GTE's workers.

Sealing of all files and destruction of all depositions.

Agreement to the terms must be unanimous; nobody gets any compensation if a single plaintiff refuses to sign.

Rocked by GTE's last-minute tactic of requesting seventy-nine new experts and by the terms of their settlement counteroffer, Rohr and Jatco buckled down to the grim realities of negotiating with a wealthy corporation and its insurance carriers. Having interpreted Carlos Martinez's gentlemanly demeanor as respect for her case, Rohr fumed: "I have lost my innocence on this case. When I see Carlos, I'll ask him how much they paid him for his integrity." She was confusing Martinez with the corporate decision makers of GTE, who had denied her the chance to attend Amy Cordova Romero's funeral.

On May 12, San Antonio trial lawyer Pat Maloney came to meet the plaintiffs for the first time, to explain to them how he was going to lead the products liability suits against Dow, Du Pont, Shell, and others. Rohr assembled the plaintiffs en masse for the first time to meet Maloney and to discuss GTE's offer to settle. Many had not seen each other for years. It was an auspicious and electric gathering.

Rohr opened the meeting by announcing that Yolanda Lozano, the plaintiff spokesperson, had suddenly deteriorated from a rapidly spreading melanoma, a cancer that had gone from her face to her brain, neck, and chest. Gasps went around the room. Rohr asked for volunteers to help her brother care for her by going to cook for her one day each. Divorced, off work for over a year, and with no insurance or family support save for her brother, Yolanda was an eerie echo of Amy Cordova Romero three years before.

Rohr then introduced Maloney, a courtly counselor who spoke with a tone and cadence unfamiliar to the New Mexicans, more like a Texas deacon or preacher than a lawyer:

> I want you all to know how lucky, lucky, lucky you are to have Josephine Rohr representing you. She is the most generous, dedicated practitioner I've ever seen. . . . We've gotten into a crusade, a holy war. Not only a just cause, but a vitally important one for workers all over this country. . . . I am filing tomorrow a suit in federal court in Houston

against the manufacturers of the chemicals that harmed you. It is thirty pages long. We are filing it in Houston because the corporations have major offices there, and because Houston, where people know about the dangers of chemicals, is where we can get you the maximum result of your suit. New Mexico ranks forty-eighth among the states in size of awards to plaintiffs like yourselves; Houston ranks first. . . . As Alan Levin has told us, do you realize that hundreds or thousands of people who come after you may suffer if we do not prevail in this effort? It puts you [*he gestured at the crowd arrayed in curved rows of chairs in front of him*] on a plane to stand for something.

I am thirty-eight years in law. I do nothing but trial lawsuits. Multimillion-dollar verdicts have been a part of our practice for a long, long time. My wife is a lawyer. My five children are lawyers, and one of them is married to a lawyer. So we bring a strong firm to bear on your good lawsuit. . . . I wouldn't dare tell you exactly how many millions of dollars we're going for. The workers' compensation cases are entirely unrelated to these new suits, so those of you dismissed with prejudice from those proceedings are not dismissed from this new suit. . . .

I love you all. We'll need a long drink of water, a stout horse, and a strong stick. And be mean as a son of a bitch. Thank you.

Maloney's son, Pat Jr., explained that the first group of products liability suits would be the group of female reproductive cancers from Department 320, followed by a group of all the other cancers; then a group of plaintiffs sharing heart and arterial problems; then plaintiffs with thyroid anomalies; and last, those suffering primarily from neurological complaints.

Free of the state law's narrow definition of occupational disease because these were separate suits in a separate jurisdiction, Josephine appealed to the plaintiffs to tell all former GTE workers they knew to come in to the office, "because with products liability, numbers help." Rohr's local associate in the case took some forty clients to another meeting room, where he counseled them to take GTE's offer. Rohr and Jatco were simultaneously telling the other half of the plaintiffs *not* to accept the offer. Plaintiffs in that first group assembled at Rohr's office later that night, asking her to represent them personally. The next day, all in the first group signed a letter demanding that the associate be removed as their lawyer. By May 19, he officially withdrew from the case.

A Last Minute Plaintiff Medical Expert

Rohr and Jatco prepared a counteroffer for GTE but waited to deliver it until the defense deposed the bioethicist, immunologist, and toxicologist Marc Lappé, who appeared as a toxicology expert for the plaintiffs. He was brought into the case by Jim Riley's Boston associate Stanley Eller, who had worked with Lappé on the Woburn case. Lappé, trained at the University of Pennsylvania in experimental pathology in the days before degrees in toxicology were offered, became active in the ethical analysis of biological research during Berkeley demonstrations against the Vietnam War in the late 1960s. After spending most of the 1970s at the Hastings Center for Bioethical Research in New York, Lappé returned to California and founded the Hazard Evaluation System and Information Service within the state Department of Health Services and Department of Industrial Relations, becoming its staff toxicologist. Lappé and his staff wrote hazard alerts about Agent Orange, plastic pipe, herbicides, and pesticides. At the time of his deposition in the GTE case, Lappé was director of the Humanistic Studies program at the University of Illinois at Chicago, an interdisciplinary program producing curricula for medical and law schools, with joint faculty appointments in the College of Pharmacy, School of Public Health, and School of Medicine. A semiotician, or reader of cultural signs, might have guessed his politics when he arrived at the Albuquerque airport wearing a blue work shirt with sleeves rolled up, faded jeans, a handlebar moustache, and Birkenstock sandals.

Lappé examined no plaintiffs and referred to none in his deposition. He gave a careful deposition, adroitly explaining toxicological concepts and limiting himself to assertions that:

Many of the chemicals used at GTE are known to be toxic. There are five published studies detailing harm to workers exposed *below* the established threshold limit value (TLV) levels of certain chemicals.

There are certain symptoms caused only by a specific chemical—for example, damage to the trigeminal facial nerve by trichloroethylene, which causes facial tics, numbness, and itching or tingling like that experienced by a number of plaintiffs.

The latest bulletin of NIOSH, a division of the federal Centers for Disease Control in Atlanta, published only weeks before Lappé gave his deposition, warns that synergistic effects of exposure to multiple chemicals lower the safe threshold of each. The bulletin refers to four different studies of worker populations, numbering between 77 and 102 individuals each, who displayed symptoms very similar to the GTE plaintiffs.[5]

Lappé testified, "It makes some difference to me from an ethical point of view that people who allegedly have experienced adverse effects from workplace exposures generally are disproportionately nonwhite." He defended the plaintiffs' claim that smelling the chemicals made them worry about harm, a connection defense experts had dismissed as nonscientific:

> I want to make very clear that this issue of the odor characteristics and warning characteristics of chemicals is now, I believe, widely, if not universally, acknowledged to be both valid and important, and critical in determining possible overexposure. As the director of the [California] Hazard Evaluation System, we sponsored research to assist in developing the first critical list of 214 chemicals and their warning characteristics. For example, trichloroethylene has an odor threshold of 23 parts per million for all but 20 percent of the population, but its TLV is 28 ppm. Therefore, to smell TCE is to be overexposed to TCE.

Lappé's most convincing testimony involved the similarity of the neurological symptoms described by GTE plaintiffs to those reported in studies cited in the recent NIOSH bulletin:

> I've reviewed standard medical texts on the various causes of polyneuropathies in a differential diagnosis. I've reviewed symptomatology to determine whether the reported prevalence of these symptoms could be found in another

population that had been exposed to viral illness or to another environmental cause outside the workplace. The overall prevalence and distribution of symptoms, even if they're off by a factor of two or more in their accuracy, are precisely the pattern of distribution in terms of commonality of symptoms that you would expect in a worker population that was exposed to organic solvents of the kind found at GTE; the skin sensitizers including, but not limited to, the epoxies of GTE; and the other chemicals we've listed.

Rohr reported that Carlos Martinez was more subdued than she had ever seen him following Lappé's testimony. Trial was less than three weeks away.

On June 1 the plaintiff team made a second settlement offer to GTE totalling $4.7 million and covering 115 workers, which Jatco characterized as "asking triple but willing to take double" the defendant's counteroffer. They asked for 60 percent of the compensation formula; a $1 million fund for plaintiffs' medical expenses, past and future; $88,500 for experts' fees; and $250,000 in attorney's fees. Their rationalization of expert fees was to allow for one testimony per plaintiff, at $750 each. For the medical fund, they figured that, rounding off the number of plaintiffs to 100 and subtracting past-due bills, the $1 million allowed less than $2,000 per plaintiff annually for five years. Their counteroffer accepted all of GTE's other terms except for two—they reserved the right to use documents gathered against GTE in preparation for product liability suits against the chemicals' manufacturers, and claimed for the plaintiffs the right to keep their own depositions and medical records.

Later in the day, Yolanda Lozano, age forty-two, died.

When Josephine hurried to the Cancer Center, she felt a shudder of *déjà vu* upon finding that Yolanda had died in room 531, the same room Josephine's son Alan had died in five years before. Later that night, Josephine called me; I listened as she talked for over two hours, grief and exhaustion driving her to try to make sense of the events of the last week, months, years.

Judge Smith postponed the trial for several days so that Rohr and the plaintiffs could attend Yolanda's funeral. Nearly one hundred plaintiffs were there; Carlos Martinez sent personal condolences.

Finally: A Deal Is Struck

After another exchange of proposals to fine tune the settlement, both sides verbally agreed on terms. Then, with Judge Smith postponing the trial day by day, Rohr had to call in all 115 of the included plaintiffs to explain the settlement offer to them and their spouses and get unanimous consent. Finally, all signed. Josephine, Rob, and Claudia took the carload of boxes of signed settlement agreements to the courthouse. She and Martinez signed the cover documents before the bench. Judge Woody Smith accepted them and signed the cover documents himself. Then he asked Rohr and Martinez into his chambers, took out a bottle of tequila, poured three shots, and toasted them.

Rohr was numb for forty-eight hours. When she returned to her office, she found that stones had shattered the windows. She was wracked with guilt that she and Jatco had been beaten, getting very little for their clients. The GTE-written press release said:

GTE Communications Systems and 115 former employees have agreed to settle claims brought against the company pursuant to the New Mexico Occupational Disease
and Disablement Act. The company and the former employees agreed not to disclose the details of the settlement,
except to say that the settlement precludes any further
action on the claims outlined in the plaintiffs' suits.

Josephine Rohr, the plaintiffs' attorney, said, "We are
pleased with the settlement, and we are pleased that GTE
has agreed to help its former employees."

Carlos Martinez, the attorney representing GTE and
its insurance carrier, said, "Although we firmly believe the
plaintiffs' illnesses are in no way related to their work at
the former GTE plant, we believe that the settlement is in
everyone's best interest.

"Continuing the legal action would have been difficult for the plaintiffs and their families, time-consuming
and costly for GTE, and divisive for the community as
a whole."

Jatco urged Rohr to see that they had done the best they
could to gain something for her workers in a case she had been
warned was impossible to win. A week after the settlement,
Judge Smith invited her and Martinez to the Rio Grande
Yacht Club for a farewell drink, and told them he would have
ruled in the plaintiffs' favor in most cases but suspected the
state supreme court would have reversed him on the many
cases filed after the three-year statute of limitations had expired, which ran from the time the worker had known her
illness was work related.

Many of the plaintiffs were confused by the settlement or

bitter about it, especially the injunction not to reveal even to family members the amount of their awards. Not surprisingly, some talked with family and friends anyway. Within a week, the *Albuquerque Journal* reported a persistent rumor that the settlement had amounted to between $2 and $3 million. Considering the trajectory of the offers and counteroffers, that is probably an accurate estimate.

One plaintiff's husband, after having the settlement explained to him, muttered to his wife, "Jesus, you got disabled twice." Ellen Kayser's son found her a 1980 tan and brown Cadillac coupe in excellent condition for $3,500, and with some of her settlement money she bought it for cash. She repaired the roof of her house and invited her children and their families for dinner. Though her health continued to deteriorate, she came by Josephine's office beaming, to show off her new car.

Rohr's share of the attorney's fees allowed her to pay off the mortgages on her house, pay back the loan she took out in 1986 to live on, replace the patched and peeling stucco on the front of her house, pay Rob and Claudia partial wages for their two years of unpaid work on the case, and take Claudia on a trip to France, where she would spend a year studying law at the Sorbonne on a scholarship. The money was soon gone.

A Separate Labor Department Settlement

The Labor Department cases filed by feisty Lee Leyba were finally settled in 1989. According to documents obtained under the Freedom of Information Act, Siemens Transmission Systems negotiated with the OFCCP a consent decree, which was presented in December 1988 to the Office of Administrative Law Judges in the Labor Department in Washington. Siemens was represented by a New Jersey law firm, and OFCCP was represented by the Labor Department solicitor's office. The consent decree details a settlement for twelve of the thirty-eight GTE workers OFCCP had identified, in its 1981 investigation, as having been made ill by conditions of their work, put on involuntary, unpaid medical leave, or fired.

The twelve workers received partial back wages. Loretto Herrera, of the PC lab tank scrubbing, got $15,000; Mercy Chavez, $8,000. Other workers received back wages in amounts ranging from approximately $11,000 to $30,000. The largest award was $49,800, paid to a man with epilepsy who twice fell into melted solder when in seizure, the second time after his doctor had explained to GTE why he fell into solder the first time. In addition, five workers received lump-sum payments for unspecified reasons. Four of these ranged from $2,000 to $8,000; a worker denied retirement benefits by her dismissal was awarded $19,220.

Leyba fumed that the solicitor's office had let Siemens's insurers pay very low totals and had negotiated away many worthy claimants, not only in the original group of thirty-eight, but others that have come to light since. The solicitor's office in Washington exempted the full OFCCP file from release under the Freedom of Information Act, so their cases may never be independently evaluated.

Suing the Chemical Manufacturers: Dow, Du Pont, and Shell

After the settlement with GTE Lenkurt, Pat Maloney, from his offices in San Antonio, took over direction of suits against the makers of the chemicals used at GTE. Josephine Rohr's law practice began to return to normal, though she continued as the emotional hub of the growing plaintiff group. By 1990, there were 277 clients involved with the chemical suits, which were filed in 151st District Court, Harris County (Houston), Texas.

In charging the chemical conglomerates with causing harm to the workers by distributing dangerous products with insufficient warning, Rohr and Maloney entered a significant battleground in law today, involving hotly contested themes and values: the limits of corporate and governmental power, the reach of justice, the meaning of social responsibility. This is the field of product liability tort law, often called "toxic torts" because of the prominent role played by poisonous substances in the era of pollution awareness. In toxic torts, business and corporate interests are arrayed against labor and consumer movements, and institutions are improvising while waiting for the paradigm to shift.

Tort Law: Unintentional Wrongs

Product liability is a new application of old tort law. Torts, originating three hundred years ago in English common law, are acts of harm for which restitution is sought in civil, not criminal, proceedings. The object of tort law, traditionally, was to restore the condition of the victim, not punish the perpetrator. Thus tort law is based on ideas of compensation, not vengeance.

The development of tort law is connected to the rise of capitalism and industrialism, which put people in positions of vulnerability to strangers who had no family or community-based ties to them and thus less incentive to protect them. At the same time, the growth of machine processes in industrial capitalism also gave the owning classes unprecedented ability to do more extensive damage through indifference or miscalculation. Tort law established that, theoretically, at least, a worker could sue his master. Ruling classes swiftly put barriers in place that dissuaded most from doing so successfully, but tort law laid the foundations of modern product liability law.[1]

Since World War II, industry has created a myriad of products and processes that can harm people, whether they are workers or consumers. Our popular horror and science fiction genres of films and writing are good mirrors of our anxiety in facing the gamut of toxins churned out in recent decades, from atomic power to DDT. With increasing rapidity, lawyers have been applying tort principles to the harms done by toxic products.[2] Judges and juries have been deciding that a manufacturer's duty to warn about dangerous products extends to all users and consumers, including the common worker in the shop or the consumer in the field. Landmark cases have been won against DES (a miscarriage-preventing drug), the

Dalkon Shield intrauterine contraceptive, and Thalidomide, a morning-sickness preventive that caused birth defects.

TORT'S POPULIST CHANGES   Over the last twenty years, judges have allowed more and more lenient definitions of the admissible causes of harm, including the introduction of "partial causation." Judges have also gradually reversed the burden of proof; plaintiffs used to have to prove beyond a shadow of a doubt that they had been harmed by a product, but now defendants must prove that they did no harm. Judges have continually, over the past thirty years, added types of injury worthy of compensation, and have steadily upped the totals of compensatory and punitive damages awarded.[3] (Many very large awards have been subsequently reduced by the trial judge or overturned on appeal.) Before this revolution, tort law made little distinction between malice, negligence, and accident; tort law was seen as remedy for all-too-human lapses.

The new tort law, however, sees harm *built into* the system of business, and seeks to strengthen public control over large-scale activities. It calls, in fact, for redistribution of social power and changes in values. The courts have been arguing that the widespread damage done with toxic materials is caused not by human error, but by deliberate policies of businesses that tailor safety investment to profit margins. These changes in tort have given it a "public law" philosophy, replacing its former corporate bias.[4] Tort keeps alive in the industrial age an earlier concept from English common law: that opposing litigants be considered equals in the eyes of the court. Faced with nearly absolute differences in power between multinational corporations and private persons, judges in the U.S. have been making efforts to level the playing field. Legal analysts recognize this emergence of product liability toxic torts as "the single most dramatic and revolutionary development in all of occupational safety and health law of the [1970s]."[5]

All concerned are asking a lot of tort law. As an effort of last resort for people suffering injuries not yet commonly understood by medicine, let alone by laymen, toxic tort suits crackle with emotions and agendas frustrated by deficiencies in other institutions breaking down under technological innovation: the sense of caring community, employer loyalty, workers' compensation, and medical primary care. In the massive Agent Orange suits, 244,000 plaintiffs represented by 1,500 law firms sued seven chemical companies, including Dow Chemical. Much as in the original GTE Lenkurt occupational disease cases, these Agent Orange plaintiffs didn't understand the legal and economic subtleties that kept the lawyers and paralegals busy in their meetings. For the plaintiffs, Peter Schuck writes,

it was less policy and social control of toxics than corrective justice, the traditional promise of tort law. The veterans saw the trial as their chance to settle accounts, to recover from the chemical companies what the war took from them—their youth, their vigor, their future. The case came to symbolize their most human commitments and passions—*insistence upon respect and recognition, hope for redemption and renewal, hunger for vindication and vengeance* [emphasis added]. For them, it was a searing morality play.[6]

The Vietnam vets got an out-of-court settlement from seven corporate defendants, including Dow Chemical, of $180 million, which has not been distributed pending appeals. It sounds like a lot, but $180 million, distributed equally to 244,000 plaintiffs, will yield about $737 per person.

THE BUSINESS COMMUNITY'S "LITIGATION CRISIS"   For supporters of fewer restrictions on business, the Rohr-Maloney suits against Dow, Du Pont, and Shell will be more evidence of an avalanche of bogus lawsuits tying the hands of American innovators. For the last decade, corporate cries to reform tort law because of a "litigation crisis" have become "a titanic struggle" waged with "incredible passion and resources."[7] The tort reform bill before Congress in 1990, Senate Bill 1400, is opposed by a coalition of consumer, labor, women's, environmental, senior citizen, and victims' rights groups, who point out that the business lobby has changed its justification for reforming tort laws every time a new bill is introduced. Victor Schwartz, lead lobbyist for the industry effort, admits, "In 1981 it was fairness; in 1987, insurance; now [in 1990 it is] competitiveness."[8]

There is, in fact, no evidence of a significant increase in product liability suits. The General Accounting Office (GAO) studied tort filings between 1981 and 1986—the height of the tort reform movement—and found that filings for product liability suits, not counting asbestos, Bendectin, and the Dalkon Shield cases, grew by about 4 percent per year, compared to 6 percent per year for all civil filings and 5 percent per year for expenditures on personal goods. Nor have awards been exploding: a 1989 GAO study of jurisdictions in five states found that plaintiffs won in fewer than 50 percent of the cases, that the highest awards went to the most severely injured plaintiffs, and that severely injured victims frequently receive less than their out-of-pocket expenses.[9]

Senate Bill 1400 would make it virtually impossible to recover punitive damages against makers of defective products. Punitive damages are already "exceedingly rare," and
are clustered in cases involving intentional conduct, not product liability.[10] Nevertheless, threat of tort judgments is the
strongest weapon workers and consumers have, and in the rare
instances in which they are awarded, punitive damages have
often led to the redesign, recall, and removal of dangerous
products from the market. Pam Gilbert, legislative director
of Public Citizen's Congress Watch and one of the leading
opponents of tort reform, says,

> Manufacturers hate to get sued, and hate to reorganize
> business practices so as to avoid harming workers, which
> is why we like product liability. Tort is a much more level
> playing field than the legislative process; business doesn't
> like taking its chances in open court, where judge and jury
> weight the believability of two stories. There is a wave
> throughout the U.S. in every area of law to cut back on
> plaintiffs' rights—in accounting disputes, malpractice, and
> product liability.[11]

Pat Maloney's Claims

This was the legal ferment into which Pat
Maloney stepped when he filed product liability suits against
Dow, Du Pont, and Shell Chemical Division in 1987. He
asked for $2.5 million in actual damages per worker, $5 million in exemplary damages per worker, and unspecified punitive damages per worker. Exemplary damages are awarded
against a defendant whose conduct the court deems to have

been poor, to make an example to others. Punitive damages are awarded if the court finds the defendant's conduct to have been shocking to the conscience. Maloney operated on the formula that if the case were settled out of court, plaintiffs got 60 percent of the award and attorneys 40 percent. If the case required a trial, the split would be 50–50. In this case, Maloney figured that settlement would be for about 30 percent of those figures, or $1 million in actual damages per plaintiff, and possibly $2 million each in exemplary damages. Charlie Nicholson, a Maloney associate, presented an outline of the case to a conference on toxic tort litigation held in New Jersey in February 1989. A panel of tort experts there reportedly assessed the case's potential at $500 million.[12]

Throughout 1988 Dow, Du Pont, and Shell worked through Du Pont's law firm. They argued that the cases should not be filed in Texas, and lost. They argued that the cases should not be filed in federal court, and won. They argued that all the plaintiffs should be dismissed because the statute of limitations for filing notice of occupational disease had expired, and lost. They argued that the plaintiffs should post a huge bond to cover the defendants' expenses, because the suits were "nuisance" suits. They lost. After over a year of jurisdictional and procedural jockeying, the defendants argued that they needed to bring in a large New York law firm to assist them. The judge ruled that the plaintiffs could bring in reinforcements also. Maloney said he would bring in Josephine Rohr; the defendants retracted their request. Maloney had already associated with Roger Christ, the attorney who had handled the plaintiffs' cases against NASA in the 1986 space shuttle *Challenger* explosion case.

During 1988, plaintiff medical expert Marc Lappé was said to have discovered that Dow and Du Pont had been the principal sources of tests supplied to the American Industrial Hygienists, who set the approved safe levels of organic solvents based on those tests done by the manufacturers. The plaintiffs also charged that Dow and Du Pont had been aware of the 1961 Swedish studies implicating solvents in brain damage, and had misrepresented English studies to justify exposure levels ten times higher. Jack Lacy, the former GTE safety officer, was also on retainer to Maloney, and he came up with a valuable source of chemical information: Canada's OSHA supplies workplaces with Material Safety Data Sheets up to five times as detailed as any seen in the United States. These dossiers on hundreds of widely used solvents and other chemicals come on compact disks, put out by the Canadian Centre for Occupational Health and Safety. Their series, called "INFODisc," requires a compact disk reader with converter for display on a personal computer. At $700–$900, the package may be too expensive for the smallest companies but well within the budget of any library, municipal entity, or major company. The Canadian information listed a number of the solvent intoxication effects on cognition, balance, and emotions that U.S. MSDSs do not mention.

The cases threatened to overwhelm the Houston state courts, so the judge appointed a special master to administer the rest of the discovery process. Maloney asked Du Pont to submit a list of past suits against the company for damages caused by toxic chemicals. Du Pont refused and lost. The master ordered Du Pont to supply Maloney within forty-five days with names of all lawsuits, workers' compensation claims, and other claims against any of the defendant firms.

Maloney and Rohr felt a surge of optimism over the fact that Du Pont's legal team seemed to be sleepwalking through the early rounds of the case, not only losing a string of pro-

cedural rulings but taking the workers' conditions lightly. By
the middle of 1990, the defendants apparently had no medical
experts retained. Maloney had offered to conduct depositions
of both sides' experts throughout 1989, but Du Pont declined
to depose Levin and Lappé and had yet to name anyone on
its own behalf. Maloney and Rohr interpreted these facts as
suggesting that Du Pont, with an estimated $1 billion in in-
surance coverage, was certain that it could settle the cases
painlessly whenever they became too irritating.

In September 1990 lawyers for Du Pont and Shell settled
with some 200 to 250 ex–GTE Lenkurt workers in another
secret settlement. A rumor made the rounds in Albuquerque
that the awards to each person were not close to the million-
dollar level Pat Maloney expected, but closer to the level paid
in the original GTE settlement, in the low to medium five-
figure range.

The Most Recent Studies of the GTE Workers

While the product liability suits were being filed,
teams of occupational health specialists who had become
aware of the GTE plaintiffs before the earlier settlement pro-
ceeded to conduct studies of the effects of chemicals on the
workers. In February 1988, the team of neuropsychologist
Rosemarie Bowler, from San Francisco, and neurophysiolo-
gist Donna Mergler, from the University of Quebec at Montreal, contacted Josephine Rohr. They went to Albuquerque
on their own time with two graduate students and evaluated
color vision and personality profiles. The results showed serious problems among the 125 former GTE workers evaluated,
and Bowler and Mergler proposed to Rohr that they return

to do a full study investigating neuropsychological and neurophysiological effects. Maloney agreed to pay for a suite of offices adjacent to Rohr's, and for lodging and food for the research team.

Mergler and Bowler asked Jim Cone and Bob Harrison, the Bay area occupational health specialists already familiar with the case, to help carry out a major study. They administered questionnaires on diagnosed illnesses, symptoms, and reproductive outcomes, and administered a battery of neuropsychological, neurophysiological, and personality assessment tests. They asked Rohr to have as many plaintiffs as possible find a "control" person matching as closely as possible their age, marital and pregnancy status, ethnicity, and health condition prior to their GTE employment.

In June 1988 a team of thirty people from San Francisco and Montreal, including clinical psychologists, biologists, physicians, and support personnel, came to Albuquerque. Mergler's colleagues were from the University of Quebec at Montreal's Center for Action-Research in Occupational Health. All came on an unpaid, volunteer basis; some used vacation time. They tested 180 former GTE workers.

The British Journal of Industrial Medicine has printed an article by Mergler, Bowler, and epidemiologist Guy Huel showing that GTE workers had four times the risk of spontaneous abortions after going to work at GTE, compared to matched controls and to their own lives before beginning work at the plant.[13] The study found a higher rate of spontaneous abortions among the GTE workers than a very similar survey done in 1988 of a group of women semiconductor workers at the Digital Equipment Corporation plant in Hudson, Massachusetts.[14]

An article by Bowler, Mergler, Rauch, Harrison, and Cone has been accepted by the *Journal of Clinical Psychology*, an official journal of the American Psychological Association. This article substantiates what the GTE workers and their families had been saying—that the workers were experiencing behavioral changes from the effects of the chemicals. Another article based on the neuropsychological testing is appearing in the journal *Neurotoxicity*. Bowler, Mergler, and Huel are also analyzing the incidence of hysterectomies, and say they find similar excess risk of that procedure among the GTE workers. As of 1988, before scores more became plaintiffs, medical records show that the number of women who had hysterectomies while employed in GTE Lenkurt assembly departments stood at 51.

Mergler, Bowler, Harrison, and Cone also collaborated on a study of vision effects from solvent overexposure. This study considers the optic nerve an "indicator" of damage to the overall neurological system. They tested a group of 137 former GTE workers and 102 matched controls, finding significantly lowered visual functions among the GTE workers as compared to the control group. There was a strong correlation between workers' reported symptoms of vision problems and measurable vision dysfunctions on a variety of tests. Forty-six percent of the workers studied reported blurry vision that was not correctable with glasses. This uncorrectable blurring is suspected to be a result of damaged retinas and optic nerves. The visual blurring in these workers was significantly related to their levels of acquired color vision loss, an indicator of retinal or optic nerve damage. The workers tested in this study worked an average of 6.3 years around solvents and other neurotoxins at GTE, and had ceased that work an average of 6.3 years before the study. In the meantime, two plaintiffs, Ellen Kayser and Grace Wessel, have gone blind in patterns that Mergler says reflect the solvent damage effects.

Reviewers at some professional journals have been critical of the Bowler-Mergler team's research articles on the grounds that they were not done with randomly selected workers and controls. Mergler defends her team's unpaid work:

> I get angry when we get that kind of criticism. Sure, they are not full-scale studies with all the controls known to medical research built into them. Sure, we tested a select population. But *none have ever been done* with these workers and with these problems. No one has funded it or allowed it. The studies we did were the best possible under the circumstances, and *they had to be done to get the ball rolling.* Someone is going to have to devote the resources to these problems that they deserve. Women around the world are making these electronic things for us. We are trying to call the research community's attention to the legitimacy of this work.[15]

On December 14, 1990, Pat Maloney won a jury trial in Houston, defeating Dow Chemical Corporation's effort to have the product liability trial summarily dismissed. Dow argued that the law required the plaintiffs to have filed their cases sooner than they did. Trial on the merits of the case is set for March 4, 1991. Among the information that may emerge in the trial, and that Dow was unsuccessful in keeping from the plaintiffs, is the number of similar suits filed against Dow in the past, both tried in court and settled, and the number of workers' compensation cases filed against the company for occupational illnesses. Maloney will thus go to trial—or into settlement negotiations again—armed with more quantifiable evidence, including the epidemiologic studies coming from the voluntary research of Bowler, Mergler, and their colleagues. Their findings will augment the testimony of Alan Levin and Marc Lappé, who remain the plaintiffs' principal expert witnesses. Most important, since it has already been published, Bowler and Mergler's work cannot be muzzled by an out-of-court settlement.

Conclusions

The evidence on this point is clear . . . institutions, both
public and private, exist because the people want them,
believe in them, or at least are willing to tolerate them.
The day has passed when business was a private matter—if it ever really was. In a business society, every
act of business has social consequences and may arouse
public interest. Every time business hires, builds, sells,
or buys, it is acting for the . . . people as well as for itself,
and it must be prepared to accept full responsibility for
its acts.
—Robert Wood Johnson, founder of Johnson and
Johnson, Inc., 1947[1]

At the beginning of this book, I described a scene of community ritual—the welcoming of a new member into a city's
life, with mutual pledging of admiration, support, and gratitude between GTE Lenkurt president Charlton Hunter and a
number of civic leaders of Albuquerque. Certainly the Albuquerque leaders were not a true cross-section of the community; the facets of community life not present at that cere-
monial dinner were nearly innumerable. Be that as it may,
the ritual of welcome and congratulations embraced a collective vision of the public/private partnership. In that sense,
the opening scene appears to suggest the potential for sowing
the seeds of public responsibility spoken of by Robert Wood
Johnson in his statement above.

By the end of the book, however, I am able to describe only
the steps in a legal process that has privatized the relation-

ship between the employer and hundreds of its workers. No community action is possible when each worker's health and well-being is judged separately, in isolation from work and from the experiences of her peers. No collective vision can chart a course into the future when autocratic management sees workers more as if they were pieces of machinery bolted to their work stations than messengers whose relationships crisscross the webs of family, neighborhood, and city life. By the end of the book, the care and gratitude expressed in the opening scene, which might have been rebestowed twenty years later, with real feeling, on many of those women who could rightfully have believed they had earned someone's respect and praise, had been transmuted into money, bestowed with reluctance and conditions. Ellen Kayser recalls visiting her former GTE Lenkurt workplace, now Siemens, after the first settlement, and speaking with one of the executives who knew her in the old days. "He spoke to me with this honeyed, patronizing tone," she said, "as if I was a little child."

Many in America are striving for the sense of community and public responsibility embodied in Robert Wood Johnson's manifesto. The GTE Lenkurt case, like many other class-action legal issues and toxic tort product liability suits, is a sign of the struggle by many to transcend the powerful forces in our society that atomize and trivialize the bonds of community. Even within the "public law" forms of recent tort law, however, we find opposing movements limiting the power of ordinary people to make connections with each other.

One of these is the trend toward court secrecy embodied in sealed civil settlements. Although judges helped plaintiffs by liberalizing tort procedures, they are now helping corporate defendants by agreeing to sealed settlements. In a four-part investigative series in 1988, *Washington Post* reporters Elsa Walsh and Benjamin Weiser found:

A system of private justice has evolved within the public courts, allowing important disputes that often involve serious questions of public safety to be resolved in secret. . . . The system has become pervasive. In local and federal courthouses across the country, there are confidentiality orders in hundreds of cases that allege safety problems with widely used products and facilities. Every day, someone gets into a car, takes a drug, sees a doctor or wakes up near a toxic site that has been the subject of a lawsuit covered by a confidentiality order. . . . Once used exclusively for cases involving business trade secrets, national security or personal privacy, [sealed settlements] are increasingly being used to prevent debate about critical problems of public safety and policy.[2]

One of the disturbing facts about the structure of this trend, developing over the last fifteen years, is its ad hoc quality. After interviewing hundreds of lawyers and judges and reading thousands of pages of rulings, Walsh and Weiser found that "there is a striking lack of consistency and standards" in the way various judges handle requests to seal files in civil cases, and "this informal approach conflicts with the long-accepted American tradition that the public has a right to see basic records in a civil lawsuit, an expectation formally upheld by the U.S. Supreme Court."[3] Records have been sealed in hundreds of cases of medical and legal malpractice, toxic contamination, and product liability suits. "Secrecy is worth money," says a Washington, D.C. attorney. "No seal, no bucks."[4]

A case involving the Valeron Corporation, a subsidiary of GTE, illustrates the improvisatory quality to some sealed settlements. A group of former workers sued Valenite, a subsidiary of Valeron, for exposing them to metal dusts that appear to cause "hard-metals disease," a crippling of the heart and lungs. One worker in the case who did not agree to a sealed settlement, a black woman, was ordered by the judge to pay the court costs for all the plaintiffs, a sum of $86,000. When the woman did not pay the costs, the judge put a lien on her house for the amount.[5]

In an important precedent, a New York judge ruled on August 16, 1989, that a case he had sealed after a settlement should be opened to benefit public health. New York Supreme Court Justice Joseph G. Fritsch had acquiesced in the sealing of records when Xerox Corporation and two New York families came to settlement terms over the alleged connections between private wells Xerox admitted contaminating with trichloroethylene (TCE), and cancer and other serious conditions suffered by children of the two families drinking the well water. John P. Morgan, chief of pharmacology at City University of New York Medical School and an expert in clinical toxicology, says he found strong causal links between the amounts of TCE ingested by the children and the types of illnesses they contracted. At Judge Fritsch's order unsealing the files, Morgan said, "I have been frustrated in my inability to share my [research in the case], and provoke the needed discussion and feedback from others in the scientific community." Senator Daniel Patrick Moynihan, who pushed for unsealing of the records, agreed. "Locking away vital health and environmental data serves no one," the senator said, "and throws up roadblocks to legitimate scientific inquiry into chemical contamination." Fritsch limited his order to unsealing epidemiological and environmental data, reports, and tests, and did not authorize release of anyone's names or medical records.[6]

These developments in toxic torts illuminate a limitation

inherent in the tort idea that huge corporations and their workers are equals. With its presumptions of equal adversaries settling personal disputes, tort law wraps court testimony in the guise of a private affair, leading to the concept of mutually agreeable secret settlements. But toxic-exposure cases often have profound public health implications because of the reach of standardized corporate practices into mass markets. There is serious tension between social responsibility and the privacy conditions of the settlement trend.

The American Bar Association is beginning to criticize sealed settlements on grounds that they deny plaintiffs equal access to counsel. In 1988, the bar association began studying gag orders and settlement provisions that extract attorneys' promises never to sue the defendant again for any reason, or help anyone else so sue, or publish what they learned during pretrial discovery (as the GTE Lenkurt settlement ordered). The bar association study group concluded that provisions of that kind may deprive other plaintiffs of their right to competent counsel and amount to restraint of First Amendment rights.[7]

However, despite these disturbing trends in the practice of product liability tort law, workers and consumers have little choice but to continue to use toxic tort suits as part of their legal defense against industrial harm, because the nontort alternatives are even weaker. Legislation, regulation, and administrative enforcement have been historically weak, and were weakened even further during the 1980s. While federal safety laws and regulations were funded, at the time of OSHA's second anniversary in 1973, at a budget level that amounted to less than $2 per worker per year, state safety laws and regulations were funded with only 40 cents per worker per year.[8] It is no wonder that regulation is not effective and that no one has ever gone to jail in the United States for endangering workers' lives and health in the workplace.[9]

In a study of OSHA fines levied against businesses found endangering worker health and safety, the National Safe Workplace Institute, a nonprofit worker advocacy group, said in 1989 that a two-tiered system had developed in the United States. Only one state OSHA levied a fine of more than $100,000 during the period 1986–1989, whereas during the same period federal OSHA proposed fines of more than $100,000 against fifty-eight plants and construction sites.[10] However, the *National Law Journal* found, in reviewing federal OSHA's "megafine" policy, that attorneys for the agency and industry routinely bargained large fines down by an average of 67.5 percent over the same period. The average paid fine for a workplace fatality caused by negligence in fiscal year 1988 was $1,442.82. There is little incentive to settle fines quickly, because OSHA has no power to compel accused violators to pay and charges no interest until after all appeals are exhausted. And appeals are not heard quickly; for months in 1989, the OSHA Review Commission lacked a quorum because of vacancies.[11]

The money spent on preventing workplace injury and illness via enforcement of regulations is so low because industry claims regulatory requirements cut profits and eliminate jobs. The evidence, according to Richard Kazis, indicates otherwise. Kazis, author of *Fear at Work: Job Blackmail, Labor, and the Environment*, a book about the argument that we must choose either a clean environment or jobs, points out these examples:

> After four workers died of liver cancer at a B. F. Goodrich plant in Kentucky in 1974, OSHA issued emergency standards for vinyl chloride exposure. Rubber industry

spokesmen claimed the new standards would cost $65 to
$90 billion; the actual cost was less than $1 billion. In
1978, the U.S. Secretary of Energy claimed the beryllium
industry would be forced to shut down if forced to clean
up carcinogenic fumes and dust, which he claimed would
cost in excess of $150 billion. Actual cost: $5 billion. Environmental laws, often accused of costing jobs, have added
a net total of at least 300,000 new jobs, considering effects
of clean air and clean water laws alone.[12]

Business resentment of health and safety regulations, at the
major corporate level, cannot be shown to flow from economic
necessity. Drug corporations' support of tort reform comes at
a time when the industry's profits were up 70 percent over the
period 1987–1989.[13]

Certainly it is hard to do business, especially in the
ever-innovating telecommunications and electronics fields.
Certainly regulation makes innovation and entrepreneurship
harder. However, free enterprise can be completely free only
at the price of jettisoning the altruistic communitarian values
and responsibilities that hold us together as a society. Only
a century ago, being "incorporated" meant being entrusted
with singular duties to benefit the whole community. Alan
Trachtenburg notes:

> The word [corporation] refers to any association of indi-
> viduals bound together into a *corpus*, a body sharing a com-
> mon purpose in a common name. In the past, that purpose
> had usually been communal or religious; boroughs, guilds,
> monasteries, and bishoprics were the earliest European
> manifestations of the corporate form. . . . It was assumed,
> as it is still in nonprofit corporations, that the incorporated

body earned its charter by serving the public good. . . .
Until after the Civil War, indeed, the assumption was
widespread that a corporate charter was a privilege to be
granted only by a special act of a state legislature, and then
for purposes clearly in the public interest. Incorporation
was not yet thought of as a right available on application by
any private enterprise.[14]

This private-public stewardship ethic persisted into the twentieth century. As late as 1911 a leading businessman in Boston,
Henry Lee Higginson, said that corporate property "belongs
to the community." [15]

Like many individuals in recent decades, some corporate
boards and executives have come to view their corporate
entities as radically free from the responsibilities that come
with good citizenship. Like distorted athletes, some see the
game as played only between the lines of competition, for-
getting that cooperation underlies all fields of human success.
There are communitarian traditions in our national history
that we must not let die. As University of California sociologist
Robert Bellah and his colleagues say,

Perhaps life is not a race whose only goal is being fore-
most. . . . We will need to remember that we did not create
ourselves, that we owe what we are to the communities that
formed us, and to what Paul Tillich called "the structure
of grace in history" that made such communities possible.
We will need to see the story of our lives on this earth not
as an unbroken success but as a history of suffering as well
as joy. We will need to remember the millions of suffering
people in the world today and the millions whose suffering
in the past made our present affluence possible.[16]

The Albuquerque workers of GTE Lenkurt tell us a story that was almost silenced—locked in the technical jargon of medical records and discarded with sealed depositions. They are not representative of all American workers and they never presumed to take over management responsibility for guiding the affairs of General Telephone and Electric Corporation. They are casualties from the shifting front lines of the information age. How ironic, that the manufacture of some of the most powerful communications equipment ever made involved workers denied the power of their own voices.

And yet, the story from the GTE Lenkurt case carries the good news of a hard-fought victory for worker compensation. Occasionally, as in this case, tort law, sacrifice, and a spirit of voluntarism by legal and medical advocates can level the playing field between massive corporations and workers from the shop floor. This book has been an account of what Dennis Hayes has called a "controlled chemical catastrophe."[17] It is also an account of the unified commitment that underlies all popular change.

Notes

Introduction

1. *Albuquerque Journal*, March 23, 1972.

2. Barry Bluestone and Bennett Harrison, *The Deindustrialization of America: Plant Closings, Community Abandonment, and the Dismantling of Basic Industry* (New York: Basic Books, 1982), p. 182.

3. Ibid.

4. *Telephony*, October 31, 1977.

5. "Why GTE Is Body Snatching," *Forbes*, July 10, 1978, p. 36.

6. Ibid.

7. John S. McClenahen, "Moving GTE Off Hold," *Industry Week*, January 12, 1981, p. 67.

8. Lenny Siegel and John Markoff, *The High Cost of High Tech: The Dark Side of the Chip* (New York: Harper & Row, 1985). Dennis Hayes, *Behind the Silicon Curtain: The Seductions of Work in a Lonely Era* (Boston: South End Press, 1989).

9. Siegel and Markoff, *The High Cost of High Tech*, p. 152.

10. Robert Howard, *Brave New Workplace* (New Yor : Viking, 1985), pp. 9–10.

Chapter One

1. Calling women by their first names has been associated historically with patronizing and patriarchal attitudes. In this book, I am reflecting the habits adopted by Rohr, her staff, and the plaintiff workers, who used first names nearly exclusively when talking about and to each other. For me, this gave their conversation shades of intimacy, affection, and egalitarianism that was subversive of the corporate hierarchy they were fighting.

2. Primary sources for plaintiffs' medical conditions and work experiences include their depositions (averaging 100 pages double-spaced,

given under oath before attorneys for both sides); their medical records, supplied by their medical providers; and supplementary interviews I conducted with many. Primary sources for the development of legal claims include the documents as filed by both sides with New Mexico district court in Albuquerque; scores of hours of observation of the operation of Rohr's office; scores of hours of interviews with her and her staff, conducted throughout 1987–89; and supplementary press accounts. Rohr secured from her clients permission to use all materials to further their cases, including for this book.

3. For a sketch of the lurid and surreal thirty-one years of Trujillo's rule, see *Time* magazine, June 9, 1961, p. 26. Among book-length treatments, the best and most available in English are Robert D. Crassweller, *Trujillo: The Life and Times of a Caribbean Dictator* (New York: Macmillan, 1966); Bernard Diederich, *Trujillo: The Death of the Goat* (Boston: Little, Brown, 1978); and Germán E. Ornes, *Trujillo: Little Caesar of the Caribbean* (New York: Thomas Nelson and Sons, 1958.) Some details of Rohr's accounts of her family's struggles under Trujillo cannot be confirmed in the above sources. As historians of the era point out, however, truth is one of the first casualties under such a dictatorship. In a reign of official disinformation and terror, history survives in word-of-mouth anecdotes and personal memory.

4. According to Crassweller's and Diederich's accounts, there were three Mirabal sisters, married to leaders of groups opposing Trujillo, and all were murdered, after months of torture, in November 1960.

Chapter Two

1. History of the negotiations between GTE Lenkurt and Albuquerque is based on newspaper accounts and on interviews I conducted September 29 and 30, 1987, with four city leaders playing key roles in those negotiations. They included Harry Kinney, then chairman of the city commission, later a multiterm mayor; Ray Powell, then an executive with Sandia National Laboratories, later a Democratic candidate for governor; James Garvin, then a Sandia Laboratories executive and president of the Industrial Foundation of Albuquerque, a board created to raise money for the Albuquerque Industrial Development Service (AIDS); and Ed Jory, a banker then president of AIDS.

2. Reports of NLRB hearings and decisions are found in *NLRB Decisions* (Chicago: Commercial Clearing House). 1973: Para. 25,555,

cases 28-CA-2495, 2506, 2613, and 2670–72; 1974: Paras. 26,070, 26,294, and 26,847.

3. Kenneth Geiser, "Health Hazards of the Microelectronics Industry," *International Journal of Health Services* 16 (January 1986): 105–15.

4. "Epoxy Resins Factsheet," Project on Health and Safety in Electronics, Santa Clara Center for Occupational Safety and Health, San Jose, California, 1981.

5. Cynthia Talbot, "Toxic Substances Commonly Found in Electronics: A Guide for Health Professionals." Project on Health and Safety in Electronics, Santa Clara County Center for Occupational Safety and Health, San Jose, California, 1981.

6. Robert Howard, *Brave New Workplace* (New York: Viking, 1985), p. 148.

Chapter Four

1. *Time* magazine, September 28, 1987, p. 51.

2. Michael Van Waas, *The Multinationals' Strategy for Labor: Foreign Assembly Plants in Mexico's Border Industrialization Program* (Ann Arbor: University Microfilms, 1981).

3. Robert Howard, *Brave New Workplace* (New York: Viking, 1985).

Chapter Six

1. *Woodrow Sterling et al v. Velsicol Chemical Corp.*

2. See, for example, George D. Clayton and Florence E. Clayton, eds., *Patty's Industrial Hygiene and Toxicology*, 3rd rev. ed., vols. IIa, IIb, IIc (New York: John Wiley, 1981).

3. Lonny Shavelson, "Tales of Troubled Waters," *Hippocrates* 2 (March–April 1988): 70–77.

4. Eliot Marshall, "Woburn Case May Spark Explosion of Lawsuits," *Science* 234 (October 24, 1986): 418–20. "Immune System Theories on Trial," *Science* 234 (December 19, 1986): 1490–92.

5. Alan S. Levin and Vera S. Byars, "Environmental Illness: A Disorder of Immune Regulation," *State of the Art Reviews—Occupational Medicine* 2 (October–December 1987): 669–81; see also Vera S. Byers, Alan S. Levin, David M. Ozonoff, and R. W. Baldwin, "Associations Between Clinical Symptoms and Lymphocyte Abnormalities in

a Population with Chronic Domestic Exposure to Industrial Solvent-contaminated Domestic Water Supply and High Incidence of Laeuke-mia," *Cancer Immunotherapy* 27 (1988): 77–81.

6. Paul Brodeur, *Outrageous Misconduct: The Asbestos Industry on Trial* (New York: Pantheon, 1986).

7. Ibid., pp. 256–57.

8. Walker's deposition is the source of his testimony about the case and his qualifications.

9. Kenneth J. Rothman, *Modern Epidemiology* (Boston: Little, Brown, 1986), pp. 12–13.

Chapter Seven

1. *New Mexico Statutes 1978 Annotated*, Pamphlet 75.

2. *Holman v. Oriental Refinery*, 75 NM 52, 400 P.2d 471 (1965).

3. Lawrence M. Friedman and Jack Ladinsky, "Social Change and the Law of Industrial Accidents," *Columbia Law Review* 67 (1967): 269–82.

4. Joseph A. Page and Mary-Win O'Brien, *Bitter Wages: Ralph Nader's Study Group Report on Disease and Injury on the Job* (New York: Grossman, 1973).

5. "Organic Solvent Neurotoxicity," *NIOSH Current Intelligence Bulletin 48*, U.S. Department of Health and Human Services, Public Health Service, Centers for Disease Control, National Institute for Occupational Safety and Health, Atlanta, NIOSH Pub. 87–104.

Chapter Eight

1. Tort origins and philosophy are discussed well in Richard L. Abel, "Torts," in *The Politics of Law—A Progressive Critique*, ed. David Kairys (New York: Pantheon, 1982). See also Lawrence M. Friedman and Jack Ladinsky, "Social Change and the Law of Industrial Accidents," *Columbia Law Review* 67 (1967): 269–82.

2. The leading case was *Borel v. Fibreboard Paper Products Corp.* (1973–74), in which asbestos workers first won damages against manufacturers. See Paul Brodeur, *Expendable Americans* (New York: Viking, 1974), and *Outrageous Misconduct: The Asbestos Industry on Trial* (New York: Pantheon, 1986). See also Adeline Gordon Levine, *Love Canal: Science, Politics and People* (Albany: State University of New York Press,

1983); Peter Schuck, *Agent Orange on Trial* (Cambridge: Harvard University Press, 1986); and Chellis Glendinning, *When Technology Wounds* (New York: Pantheon, 1989).

3. Shuck, *Agent Orange on Trial.*

4. David Rosenburg, "The Causal Connection in Mass Exposure Cases: A 'Public Law' Vision of the Tort System," *Harvard Law Review* 97 (February 1984): 851.

5. Mark A. Rothstein, *Occupational Safety and Health Law*, 2d ed. (St. Paul: West Publishing, 1983).

6. Schuck, *Agent Orange on Trial*, p. 256.

7. "Product Liability Bill Provides Opportunity for Long-Term Milking of PACs by Congress," *Wall Street Journal*, June 21, 1990, p. A-16. "Is Liability Lobbying Paying Off?" Washington Update, *National Journal*, April 14, 1990, p. 907.

8. *Legal Times*, February 5, 1990, p. S-3.

9. "The 'Product Liability Reform Act' (S. 1400): Windfall for Makers of Unsafe Products, Dangerous for Consumers," Public Citizen's Congress Watch, Spring 1990. See also "Congressional Alert," *Nation's Business*, March 1988, p. 76; Marc Galanter, "Reading the Landscape of Disputes: What We Know and Don't Know (And Think We Know) About Our Allegedly Contentious and Litigious Society," *U.C.L.A. Law Review* 4 (1983), pp. 5–69; and Swartz Kindregan, "The Assault on the Captive Consumer: Emasculating the Common Law of Torts in the Name of Reform," *St. Mary's Law Journal* 18 (1987): 673.

10. "The 'FDA Excuse'; Pharmaceutical Firms Want to Blame the Government when They Place Dangerous Drugs and Medical Devices on the Market," report issued by Public Citizen's Congress Watch, U.S. Public Interest Research Group, Consumers Union, and Consumer Federation of America, May 1990, pp. 2–3. A study cited in this report finds only 4 punitive awards upheld out of 172 cases.

11. Personal communication, June 29, 1990.

12. Interview, Josephine Rohr, June 25, 1990.

13. Guy Huel, Donna Mergler, and Rosemarie Bowler, "Evidence for Adverse Reproductive Outcomes Among Women Microelectronic Assembly Workers," *British Journal of Industrial Medicine* 47 (6): 400–404 (June 1990).

14. H. Pastides, E. J. Calabrese, D. W. Hosmer, Jr., and D. R. Harris, Jr., "Spontaneous Abortion and General Illness Symptoms Among Semiconductor Manufacturers," *Journal of Occupational Medicine* 30 (1988): 543–51.

15. Personal communication, June 25, 1990.

Conclusions

1. Quoted from an excerpt of "Or Forfeit Freedom," by General Robert Wood Johnson, in the series "Public Courts, Private Justice," Benjamin Weiser and Elsa Walsh, *Washington Post*, October 25, 1988, p. A-1.

2. Walsh and Weiser, "Public Courts," October 24, 1988, p. A-20.

3. Ibid.

4. Ibid, p. A-21.

5. *Today Show*, June 26, 1990; segment produced and reported by John Alpert, a producer for Downtown Community Television, New York.

6. Benjamin Weiser, "Release of Sealed Records Ordered in Xerox Toxic-Chemical Case," *Washington Post*, August 17, 1989, p. A-24.

7. Ethics column by George Kuhlman, "The Right Choice," *ABA Journal*, January 1, 1988, p. 118, discussion of Model Rule 5.6, "Restrictions on [lawyers'] Right to Practice."

8. Joseph A. Page and Mary-Win O'Brien, *Bitter Wages: Ralph Nader's Study Group Report on Disease and Injury on the Job* (New York: Grossman, 1973).

9. *Time* magazine, "Blood, Sweat and Fears," September 28, 1987, p. 51.

10. *Chicago Tribune*, July 2, 1989.

11. Marianne Lavelle, *National Law Journal*, 1989, reprinted in *Albuquerque Journal*, January 14, 1990.

12. Richard Kazis, director of research, Environmentalists for Full Employment: testimony before Massachusetts Legislature, March 9, 1983.

13. Public Citizen's Congress Watch and others, "The 'FDA Excuse,'" pamphlet, May 1990, p. 31.

14. Alan Trachtenburg, *The Incorporation of America: Culture and Society in the Gilded Age* (New York: Hill and Wang, 1982), pp. 5–6. Quoted in Robert Bellah and others, *Habits of the Heart: Individualism and Commitment in American Life* (New York: Harper & Row, 1985.)

15. Bellah and others, *Habits of the Heart*, p. 290.

16. Ibid., p. 295.

17. Dennis Hayes, *Behind the Silicon Curtain: The Seductions of Work in a Lonely Era* (Boston: South End Press, 1989), p. 80.

Selected Bibliography

The Social Context

Baker, Robin, and Sharon Woodrow. "The Clean, Light Image of the Electronics Industry: Miracle or Mirage?" In *Double Exposure: Women's Health Hazards on the Job and at Home*, edited by Wendy Chavkin. New York: Monthly Review Press, 1984.

Bleier, Ruth. *Science and Gender: A Critique of Biology and Its Theories on Women*. New York: Pergamon Press, 1984.

Braverman, Harry. *Labor and Monopoly Capital: The Degradation of Work in the Twentieth Century*. New York: Monthly Review Press, 1975.

Chavkin, Wendy, ed. *Double Exposure: Women's Health Hazards on the Job and at Home*. New York: Monthly Review Press, 1984.

Chirot, Daniel. *Social Change in the Modern Era*. New York: Harcourt, Brace and Jovanovich, 1987.

Fernandez-Kelly, Maria-Patricia. *For We Are Sold, I and My People*. Albany: State University of New York Press, 1983.

Frobel, Folker, Jurgen Heinrichs, and Otto Kreye. *The New International Division of Labor*. New York: Cambridge University Press, 1980.

Geiser, Kenneth. "Health Hazards of the Microelectronics Industry." *International Journal of Health Services* 16 (1): 105–15 (January 1986).

Green, Susan S. "Silicon Valley's Women Workers: A Theoretical Analysis of Sex-Segregation in the Electronics Industry Labor Market." In *Women, Men, and the International Division of Labor*, edited by June Nash and Maria-Patricia Fernandez-Kelly. Albany: State University of New York Press, 1983.

Grenier, Guillermo. *Inhuman Relations*. Philadelphia: Temple University Press, 1988.

Grossman, Rachel. "Women's Place in the Integrated Circuit." *Southeast Asia Chronicle* 66:2–17 (January–February 1979).

Hembree, Diana. "Dead End in Silicon Valley." *The Progressive* 49 (10): 18–24 (October 1985).

Katz, Naomi, and David S. Kemnitzer. "Women and Work in Silicon Valley." In *My Troubles Are Going to Have Trouble with Me*, edited by Karen Brodkin Sacks and Dorothy Remy. New Brunswick, N.J.: Rutgers University Press, 1984.

Lim, Linda Y. C. "Women's Work in Multinational Electronics Factories." In *Women and Technological Change in Developing Countries*, edited by Roslyn Dauber and Melinda C. Cain. Denver: Westview Press, for the American Association for the Advancement of Science, 1981.

Lin, Vivian. "Health, Women's Work, and Industrialization: Women Workers in the Semiconductor Industry in Singapore and Malaysia." Working Paper #130, Michigan State University Women in International Development Series, 1984.

———. "Productivity First: Women's Health Under Japanese Management." *Bulletin of Concerned Asian Scholars* 16 (4): 12–25 (October–December 1984).

———. "Health and Welfare and the Labour Process: Reproduction and Compliance in the Electronics Industry in Southeast Asia." *Journal of Contemporary Asia* 16 (4): (1986).

McBride, Gail. "Fumes in the Factory" [GTE]. *The Progressive* 52 (1): 24–28 (January 24, 1988).

Noble, David F. *America by Design: Science, Technology, and the Rise of Corporate Capitalism*. New York: Oxford University Press, 1979.

Siegel, Lenny. "The Chip Is Here to Stay." *The Progressive* 49 (10): 25 (October 1985).

Siegel, Lenny, and John Markoff. *The High Cost of High Tech: The Dark Side of the Chip*. New York: Harper & Row, 1985.

Sklair, Leslie. *Assembling for Development: The Maquila Industries in Mexico and the United States*. New York: Unwyn and Hyman, 1988.

Spake, Amanda. "A New American Nightmare?" *Ms.* 14 (9): 35–36, 40–42, 93–95 (March 1986).

Stanley, Autumn. "Women Hold Up Two-Thirds of the Sky: Notes for a Revised History of Technology." In *Machina ex Dea*, edited by Joan Rothschild. New York: Pergamon Press, 1983.

Tiano, Susan. "Maquiladoras, Women's Work, and Unemployment in Northern Mexico." Working Paper #43, Michigan State University Women in International Development Series, 1984.

———. "Export Processing, Women's Work, and the Employment Program in Developing Countries: The Case of the Maquiladora Program in Northern Mexico." *Western Sociological Review* 15 (1): 53–78 (1987).

Wade, R., and M. Williams. *Semiconductor Industry Study*. California Department of Industrial Relations, Division of Occupational Safety and Health (Cal-OSHA), Task Force on the Electronics Industry, 1983.

White, Lawrence. *Human Debris: The Injured Worker in America*. New York: Seaview-Putnam, 1983.

Health Problems of Stress and Chemical Exposure

Anderson, Ronald, and others. "Access to Medical Care Among the Hispanic Population of the Southwest." *Journal of Health and Social Behavior* 22 (1): 78–89 (March 1981).

Bowler, Rosemarie, Carrie D. Thaler, and Charles E. Becker. "California Neuropsychological Screening Battery (CNS/B I & II)." *Journal of Clinical Psychology* 42 (6): 946–55 (November 1986).

Bowler, Rosemarie, and others. "Affective and Personality Disturbance of Women Former Microelectronics Workers," *Journal of Clinical Psychology* (forthcoming).

———. "Neuropsychological Impairment of Former Microelectronics Workers," *NeuroToxicology* 12:87–104 (1991).

Clayton, George D., and Florence E. Clayton, eds. *Patty's Industrial Hygiene and Toxicology*, vols. IIa, IIb, IIc. 3rd rev. ed. New York: John Wiley and Sons, 1981.

Cohen, B. G., and others. "An Investigation of Job Satisfaction Factors in an Incident of Mass Psychogenic Illness at the Workplace." *Occupational Health Nursing* 26 (1): 10–16 (January 1978).

Colligan, M. J., J. W. Pennebaker, and L. R. Murphy. *Mass Psychogenic Illness*. Hillsdale, N.J.: Lawrence Ehrlbaum Associates, 1982.

Cone, James E. "Health Hazards of Solvents." *State of the Art Reviews— Occupational Medicine* 1 (1): 69–87 (January–March 1986).

Faust, Halley S., and Lawrence B. Brilliant. "Is the Diagnosis of 'Mass Hysteria' an Excuse for Incomplete Investigation of Low-Level Environmental Contamination?" *Journal of Occupational Medicine* 23 (1): 22–26 (January 1981).

Garabrant, David H., and Robert Olin. "Carcinogens and Cancer Risks in the Microelectronics Industry." *State of the Art Reviews—Occupational Medicine* 1 (1): 119–34 (January–March 1986).

Huel, Guy, Donna Mergler, and Rosemarie Bowler. "Evidence for Adverse Reproductive Outcomes Among Women Microelectronic Assembly Workers." *British Journal of Industrial Medicine* 47 (6): 400–404 (June 1990).

Kerckoff, A. C. "A Social Psychological View of Mass Psychogenic Illness." In *Mass Psychogenic Illness*, edited by M. J. Colligan, J. W. Pennebaker, and L. R. Murphy. Hillsdale, N.J.: Lawrence Ehrlbaum Associates, 1982.

LaDou, Joseph. "Health Issues in the Microelectronics Industry." *State of the Art Reviews—Occupational Medicine* 1 (1): 1–11 (January–March 1986).

————. "The Not-So-Clean Business of Making Chips." *Technology Review* 87(4):23–36 (May–June 1984).

Levin, Alan, and Vera Byars, "Environmental Illness: A Disorder of Immune Regulation." *State of the Art Reviews—Occupational Medicine* 2 (4):669–81 (Winter 1987).

Marshall, Eliot. "Immune System Theories on Trial." *Science* 234:1490–92 (December 19, 1986).

Mergler, Donna, and Lucie Blain. "Assessing Color Vision Loss Among Solvent-Exposed Workers." *American Journal of Industrial Medicine* 12:195–203 (1987).

Mergler, Donna, and others. "Chromal Focus of Acquired Chromatic Discrimination Loss and Solvent Exposure Among Printshop Workers." *Toxicology* 49:341–48 (1988).

Mergler, Donna, Lucie Blain, and Jean-Pierre Lagace. "Solvent-Related Color Loss: An Indicator of Neural Damage?" *International Archives of Occupational and Environmental Health* 59:313–21 (1987).

"Organic Solvent Neurotoxicity." *NIOSH Current Intelligence Bulletin 48.* Atlanta: U.S. Department of Health and Human Services, Public Health Service, Centers for Disease Control, National Institute for Occupational Safety and Health, March 1987. NIOSH Pub. #87-104.

Pastides, H., E. J. Calabrese, D. W. Hosmer, Jr., and D. R. Harris, Jr. "Spontaneous Abortion and General Illness Symptoms Among Semiconductor Manufacturers." *Journal of Occupational Medicine* 30: 543–51 (1988).

Restak, Richard. *The Brain.* New York: Bantam, 1984.

Sandmeyer, Esther E. "The Aromatic Hydrocarbons." In *Patty's Industrial Toxicology*, edited by George D. Clayton and Florence E. Clayton. 3rd rev. ed. New York: John Wiley and Sons, 1981.

Schottenfeld, R. S., and M. R. Cullen. "Occupational-Induced Post-Traumatic Stress Disorders." *American Journal of Psychiatry* 142 (2): 198–202 (1985).

Selye, Hans. *The Stress of Life*. Rev. ed. New York: McGraw-Hill, 1976.

Stellman, Jeanne M., and Susan M. Daum. *Work Is Dangerous to Your Health: A Handbook of Health Hazards in the Workplace and What You Can Do About Them*. New York: Pantheon, 1973.

Tan Kok Jin and Phoon Wai Hoong. "An Episode of Trichloroethylene Poisoning." *Annals Academy of Medicine Singapore* 9 (4):532–35 (October 1980).

Van Strum, Carol. *A Bitter Fog: Herbicides and Human Rights*. San Francisco: Sierra Club Books, 1983.

Wiederholt, W. C., and others. "Clinical Ecology: A Critical Appraisal." *Western Journal of Medicine* 144:239–45 (February 1986).

Legal and Political Remedies

Abel, Richard L. "Torts." In *The Politics of Law—A Progressive Critique*, edited by David Kairys. New York: Pantheon, 1982.

Anton, Thomas J. *Occupational Safety and Health Management*. New York: McGraw-Hill, 1979.

Black, Bert, and David E. Lilienfeld. "Epidemiologic Proof in Toxic Tort Litigation." *Fordham Law Review* 52:732 (April 1984).

Brodeur, Paul. *Expendable Americans*. New York: Viking Press, 1974.

————. *Outrageous Misconduct: The Asbestos Industry on Trial*. New York: Pantheon, 1986.

Friedman, Lawrence M., and Jack Ladinsky. "Social Change and the Law of Industrial Accidents." *Columbia Law Review* 67:269–82 (1967).

Horwitz, Morton J. "The Doctrine of Objective Causation." In *The Politics of Law—A Progressive Critique*, edited by David Kairys. New York: Pantheon, 1982.

Kairys, David, ed. *The Politics of Law—A Progressive Critique*. New York: Pantheon, 1982.

Kazis, Richard. *Fear at Work: Job Blackmail, Labor, and the Environment*. New York: Pilgrim Press, 1982.

Levine, Adeline Gordon. *Love Canal: Science, Politics, and People*. Albany: State University of New York Press, 1983.

Mulcahy, Myra Paiewonsky. "Note: Proving Causation in Toxic Tort Litigation." *Hofstra Law Review* (Summer 1983): 1299.

Page, Joseph A., and Mary-Win O'Brien. *Bitter Wages: Ralph Nader's Study Group Report on Disease and Injury on the Job*. New York: Grossman Publishers, 1973.

Rahdert, Mark C. "Of Impressionists and Rohrschach Blots." Book review of *Total Justice*, by Lawrence M. Friedman. *Columbia Law Review* 86:1283 (October 1986).

Rosenberg, David. "The Causal Connection in Mass Exposure Cases: A 'Public Law' Vision of the Tort System." *Harvard Law Review* 97:851 (February 1984).

Rothstein, Mark A. *Occupational Safety and Health Law*. 2d ed. St. Paul: West Publishing, 1983.

Schuck, Peter H. *Agent Orange on Trial: Mass Toxic Disasters in the Courts*. Cambridge, Mass.: Harvard University Press, 1986.

White, Lawrence. *Human Debris: The Injured Worker in America*. New York: Seaview-Putnam, 1983.

Index

Activist networks, women's, 79–80, 86–87, 94–95, 97, 99–100

Albuquerque: city leaders recruit GTE Lenkurt, 3–4, 30–32; riot police at Lenkurt protest, 66–67

Albuquerque Industrial Development Service (AIDS), 30

Albuquerque Journal, 67, 134

Albuquerque Technical-Vocational Institute, 33

American Bar Association, opposes sealed settlements, 155

American Council of Governmental and Industrial Hygienists, 115, 144

Anaya, Diego, 43

Anaya, Mary, 62

Aragon, Patsy, 26

Assassinations, in Dominican Republic, 20–22

"Assembly line hysteria" diagnosis, 44. *See also* Hannah, Fritz; Lacy, Jack

Attorneys. *See* Eller, Stanley; Hawes, Mandy; Jatco, Roger; Maloney, Pat; Martinez, Carlos; Riley, James; Rohr, Josephine DeLeon

Banghart, Lila Navarro, 44–45

Bell system, deregulation affects GTE, 5

Bellah, Robert, 158

Bencosme family, 20–22

Blinn, Thomas, 67, 76

Bowler, Rosemarie, 100–101, 145–49

Bowling, Diane, 40–41

Brodeur, Paul, 110–11

Brophy, Theodore, 5–7

Byars, Vera, 107–8

C de Baca, Mary Lou, 52–53

California Hazard Evaluation System and Information Service, 128–30. *See also* Lappé, Marc

Campbell, Betty, 48, 73

"Car wash" degreasing machine, 41

Case studies of plaintiffs, 49–64

Caudill, Janet, 40, 45, 56–58

CBS Evening News, 116

Chavez, Mercy, 54–56, 135

Chemically-induced immune system dysregulation, 102, 106–9

Chemicals: at Advanced Micro Devices (Silicon Valley), 94; effects of, reported by GTE Lenkurt workers, 7–8, 46–47, 51; and "environmental illness," 107–8; established "safe" levels of, 115, 143–44; handling, at GTE Lenkurt, 38–42, 55, 60–63; harm from, below safe levels, 129–30; known effects of, 10–12, 50, 57, 114, 130–31

Chemicals, most common at GTE Lenkurt: acids, 36, 39–40,

Chemicals, most common (cont.):
    43; epoxies, 37–38, 49–50,
    54–55, 77; liquid plastic, 37;
    odor of, as warning signal,
    42, 130; solder, 37, 74, 135;
    solvents (freon mixtures, 1-1-
    1-trichloroethane, alcohols,
    trichloroethylene, xylene), 41–
    42, 45, 47–48, 56; synergistic
    effects of, 12, 129–30; test-
    ing of, by manufacturers, 12,
    143–44
Chip capacitors dept. *See* Depart-
    ment 341
Cohort studies, of plaintiff groups'
    health problems, 145–49
Communists, at GTE Lenkurt, 67
Communities, as source of iden-
    tity, 158
Component assembly dept. *See* De-
    partment 320
Cone, James, 100–101, 146–47
Conrad, Sheila, 87
Cordova Romero, Amy: becomes
    Rohr's first plaintiff, 17–
    19; death of, 95; history and
    medical record of, 17, 24–27
Corporate responsibility to commu-
    nity, 151–52, 157–59
Crystal filters dept. *See* Depart-
    ment 345

Deaths: Amy Cordova Romero, 95;
    Ellen Kayser's list, 62–63;
    Carlotta Leon, 119; Yolanda
    Lozano, 131–32; Jenny Mon-
    taño, 119; Isabel Romero,
    63–64; Laird Wood, 119
Defense industry, 45–46, 103
Department 320 (component as-
    sembly), 37–38, 49, 52–54,
    72, 74
Department 341 (chip capacitors), 40,
    59, 74

Department 345 (crystal filters), 36,
    40, 44, 52, 56, 74
Depositions: defined, 161n.2; impact
    of workers', on judge, 9; of
    medical experts for GTE,
    109, 111–15, 164n.8; of medi-
    cal experts for plaintiffs, 109,
    121–31
Disabled Workers United (Silicon
    Valley), 94
Doctors. *See* Byars, Vera; Cone,
    James; Harrison, Robert;
    Levin, Alan; Walker, Alexan-
    der; Wiese, William
Dominican Republic, 20–22
Dow Chemical Corporation, 116, 148
Du Pont Corporation, 116, 145

Electronics manufacturing: and
    chemicals, 11–12; criticism
    and boosting of, 14–16; social
    organization of work in, 34–41
Eller, Stanley, 128
"Environmental illness" theory,
    107–8
Environmental Protection Agency
    (EPA), 16
Environmental regulation vs. jobs
    argument, 156–57
Epidemiology Resources, Inc., 99,
    109–11
Ethnicity and gender: and abnormal
    cancers in New Mexican His-
    panics, 7–8; in plaintiff group,
    6–8, 81; and toxics exposure,
    130. *See also* Plaintiff group,
    GTE Lenkurt
Expert witnesses. *See* Medical expert
    witnesses

Fantus factory relocation service, and
    right-to-work laws, 31
Freedom of Information Act, 134–35

Frisch, Justice Joseph G., and N.Y.
State Supreme Court unsealed
settlement, 154

Garcia, Teresa, 46–47
Garvin, James, 31–32, 162n.1
Gender, of plaintiffs. *See* Ethnicity
and gender
General Electric Corporation, 30
General Telephone and Electric
Corporation (GTE), 3; elimi-
nates telephone division safety
staff, 76–77; global expan-
sion, 4–5; health problems in
subsidiaries, 153–54
Gilbert, Pam, 142
Glawe, Jackie and Millie, 96
"Good business climate," 4
GTE Lenkurt Albuquerque plant:
accidents and chemical spills
at, 40–41, 50, 59–60, 90;
defense contracts at, 45–46;
history of, 29–34, 77; management response of, to worker
complaint, 7, 44–45, 65–68,
76; medical and dispensary
programs, 52, 59–60, 72–
73, 88–89, 113; and medical
expert witnesses, 99, 109–
16; organization and conditions of work on shop floor,
35–44, 51–53, 56–57, 59–
60; press relations, 13, 87,
121, 132–33; relations with
union, 31, 34, 65–68, 70–
71; and San Carlos, Calif.,
plant, 31–32, 75–76; sealed
settlement terms, 10, 55–56,
119–21, 124–25, 134–35;
size of legal case, 8–9, 12–
13, 91–94; worker health and
work records, 29, 36–37, 88–
89, 113. *See also* Lacy, Jack;
Ventilation-exhaust

Hamilton, Dennis, 43–44, 74
Hannah, Fritz, 87
"Hard metals disease" in GTE
subsidiary, 153–54
Harrison, Robert, 100–101, 108,
146–47
Hawes, Mandy: and activist networks, 86–87, 94, 99–100;
and role of lawyers in gathering microelectronics toxics
data, 80, 97
Hayes, Dennis, 14, 159
Health insurance: ex-workers in crisis
without, 17–18, 24–27, 126;
GTE Lenkurt's two plans, 75
Health problems of plaintiffs: cancer, 17–18, 24–27, 61–64,
84–85, 119, 126, 131–32;
gastro-intestinal, 53, 56; heart,
57, 73, 119; major studies
of, by researchers, 85, 145–
49; neurological, 41, 53, 114,
129, 145–48; and one month's
log of dispensary visits, 73;
psychological, including narcosis and manic states, 39, 43,
46–48, 53, 63, 73, 81, 87, 89–
90, 100–101; and rashes, 45,
50, 55, 70, 73; in reproductive
system, 18, 51–52, 57–58, 82,
121, 145–47; respiratory, 41,
51, 53–54, 57; thyroid, 53,
60–62
Herrera, Loretto, 42–43, 135
High-tech industry, books about, 14
Hull, Harry, 84–85
Hunter, Charlton ("Chuck"), 3–4, 6,
31–32
Hysterectomies and plaintiffs, 18, 52,
57–58, 121, 147

IBM Corporation, advertising, 14–15
Illness symptoms. *See* Chemicals,

Illness symptoms (cont.):
    known effects of; Health
    problems of plaintiffs
Industrial Foundation of Albu-
    querque, 3, 30
International Brotherhood of Elec-
    trical Workers (IBEW) union:
    clashes with GTE Lenkurt
    management, 31, 34; files
    grievances with federal
    OSHA, 68–69; 1978 walkout/
    lockout and strike, 65–68

Jatco, Roger, 116–17, 133–34
Johnson, Robert Wood, 151

Kaufman, Audie, 65
Kayser, Ellen, 33, 48, 58–64, 134,
    148, 152
Keller, Maureen, 76
Kelly, Jim, 62
Kemper Insurance, 55–56, 75
Kerr, George, 51
Key, Charles, 83–84

Lacy, Jack: as federal OSHA inspec-
    tor, 71–72; as GTE Lenkurt
    safety and loss control officer,
    37, 72–77; involvement of, in
    litigation, 83, 144
Lappé, Marc, 42, 128–31, 143–44
Larkin, Juanita, 66–68
Leon, Carlotta, 86, 119
Lessin, Nancy, 100
Levin, Alan, 101–9, 127, 149
Leyba, Lee, 69–70, 134–35
Lozano, Yolanda, 33, 95, 131–32
Luna, Lupe, 62

Maloney, Pat, 117, 126–27, 137,
    142–46

Martinez, Carlos, 11, 47, 82, 87–89
Mass-COSH (Committee on Safety
    and Health), 100
Material Safety Data Sheets
    (MSDSs), about chemical
    properties, 74, 77, 144
Matta, Cecilia Flores, 45
Medical expert witnesses: chemical
    industry opposition, 102–3;
    of GTE Lenkurt, 99, 109–16,
    123; of plaintiffs, 9, 99, 101–9,
    127–31, 143–44, 149; sealed
    settlements' prior restraint
    of, 125
Medical records at GTE Lenkurt
    Albuquerque plant, 87–
    89, 113
Mergler, Donna, 145–49
Mexico, 78
Microelectronics. See Electronics
    manufacturing
Miera, Sadie, 49–52
Minority workers. See Ethnicity and
    gender
Montaño, Jenny, 119
Montaño, Vince, 90
MSDSs. See Material Safety Data
    Sheets

NASA space shuttle Challenger
    case, 143
National Institute for Occupational
    Safety and Health (NIOSH),
    16, 44, 115, 129–30
National Labor Relations Board
    (NLRB), 34, 67
National Loss Control (NATLSCO),
    73–74
Native Americans at GTE Lenkurt,
    63–64
New Mexican Hispanics. See Eth-
    nicity and gender

New Mexico State epidemiologist. *See* Hull, Harry

Occupational Disease and Disablement Act of New Mexico. *See* Workers' compensation
Occupational Safety and Health Administration (OSHA), federal: funding, enforcement, and standard-setting, 69, 155–56; and GTE Lenkurt workers, 63, 68, 78; inspector Jack Lacy, and GTE Lenkurt, 68, 71–78
Office of Federal Contract Compliance Program (OFCCP), 69–70, 134–35
Oppenheimer, Carol, 79, 85
OSHA, of New Mexico, 76–77, 79, 85

Participant observation research method, 10
Permissible exposure levels (PELs) for chemicals, 115. *See also* Threshhold limit values
Plaintiff group, GTE Lenkurt: case studies, 49–64; demographics, 7, 9, 49, 87, 122–23; as difficult for solo lawyer to manage, 19, 82–83, 87–88, 90–94, 133–34; and "double/triple shift" of work and housework, 50, 52; gender and ethnicity of, 9, 81; medical knowledge about, 52–53, 56–57, 86, 113, 145–49. *See also* Deaths
Pong, Sil, 45
Powell, Ray, 31–32
Printed circuit (PC) lab, 36, 39, 42–43, 62–63, 71
Product liability suits, against Dow,

Du Pont, and Shell, 116–17, 137, 142–45. *See also* Toxic torts
Project on Health and Safety in Electronics (PHASE), Santa Clara, Calif., 87, 89
Public Citizen's Congress Watch, and tort reform, 142

Randall, John, 38–39
Reagan administration, and OSHA enforcement record, 69
Right-to-work laws, 31
Riley, James, 101–2, 128
Rogers, Betty, 72–73
Rogers, Sam, 76–77
Rohr, Josephine DeLeon: and activist groups, 79–80, 86–87, 94, 99; in California and New Mexico, 22–23; difficulty of, fighting multinational corporation alone, 90–97; in Dominican Republic, 20–22; hit hard by deaths, 23, 95, 131–32; personal coincidence with plaintiff group, 27, 181; stages in lawsuit, 7, 17–19, 79–97, 117, 119, 132–33; and state epidemiologist, 84–85; violence against, 9–10, 132
Romero, Amalia, 86
Romero, Isabel, 62–64
Ruiz, Connie, 69–70
Rusk, David, 66–67

St. Joseph Hospital, 18, 25–26, 62–64
Sena, Mary, 33, 47
Sena, Sylvia, 86
Settlements, sealed: Du Pont, Shell, and plaintiff group (1990), 145; GTE Lenkurt, with Mercy Chavez, 55–56; GTE

Settlements, sealed (cont.):
    Lenkurt, with plaintiff group,
    10, 119–35; Labor Depart-
    ment, with GTE/Siemens
    Transmission Systems, 134–
    35; recent trends, 152–55
Shannon, Ruby, 62
Shell Chemical Division, 116, 145
Shuck, Peter, 140
Siegel, Lenny, 14
Siemens Corporation, 116, 134
"Signetics Three," 44, 47, 80
Singer-Friden Corporation, 30,
    58–59
Smith, Judge Woody, 93–94, 119–21,
    132–33

Taylor, Lynda, 80
Tena, Roberta, 86
Threshold limit values (TLVs) for
    chemicals, 115, 143–44. See
    also Permissible exposure
    levels (PELs) for chemicals
Toxic torts: against various products,
    103, 106, 110–11, 140; reform
    and "litigation crisis," 141–42,
    157; and tort philosophy and
    history, 137–40, 155
Trujillo, Rafael, 21–22, 162n.3
Tuma, Dorothy, 62
Tumor registry, of New Mexico,
    83–84

Union at GTE Lenkurt. See Interna-
    tional Brotherhood of Electri-
    cal Workers (IBEW) union
Unionization in electronics manufac-
    turing, 34

University of New Mexico medical
    institutions, 24, 83–84, 100

Vanderslice, Tom, 5–7
Velsicol Chemical Corporation, toxics
    case, 103–6
Ventilation-exhaust: experts' inspec-
    tion of, at GTE Lenkurt,
    43–44, 71, 74, 115; at Sig-
    netics in Silicon Valley, 44;
    worker problems with, at
    GTE Lenkurt, 12, 32, 36, 38,
    43–44, 50

Walker, Alexander, 109–16
Walker, Vonnie, 44
Wessel, Grace, 40, 86, 148
Wiese, William, 100
Wiggins, Jack, 51
Winkless, Dan, 35–36
Woburn, Mass., toxics case, 101
Women activist networks. See Activist
    networks, women's
Women workers, and "triple shift"
    of housework, 50–52. See also
    Ethnicity and gender
Wood, Laird, 119
Wooten, Ed, 62
Work, meaning of, in high-tech era,
    14–16
Workers' compensation, 12–13,
    55–56, 122–23
Wright, Cammie, 62

Xerox Corporation, and unsealed
    settlement, 154